DICTIONARY/OUTLINE
of Basic Statistics

BY

JOHN E. FREUND
Arizona State University

AND

FRANK J. WILLIAMS
San Francisco State University

DOVER PUBLICATIONS, INC., New York

Published in Canada by General Publishing Company, Ltd., 30 Lesmill Road, Don Mills, Toronto, Ontario.

Published in the United Kingdom by Constable and Company, Ltd., 3 The Lanchesters, 162–164 Fulham Palace Road, London W6 9ER.

This Dover edition, first published in 1991, is an unabridged, slightly corrected republication of the work first published by the McGraw-Hill Book Company, New York, 1966.

Manufactured in the United States of America

Dover Publications, Inc., 31 East 2nd Street, Mineola, N.Y. 11501

Library of Congress Cataloging-in-Publication Data

Freund, John E.

Dictionary/outline of basic statistics / by John E. Freund and Frank J. Williams.

p. cm.

Reprint. Originally published: New York : McGraw-Hill, 1966.

Includes bibliographical references.

ISBN 0-486-66796-0

1. Statistics—Dictionaries. I. Williams, Frank Jefferson. II. Title.

QA276.14.F74 1991

519.5′03—dc20 91-3467

CIP

PREFACE

This book was written for all those who, in their work or in their studies, become involved with modern statistics and its related fields—*for the specialist in the biological, social, and natural sciences* who must have some knowledge of experimental design, statistical inference, and decision theory; *for the (increasingly nonintuitive) manager* who must not only understand basic business statistics and the methods of quality control, but who must also have some familiarity with the language used in operations research, the theory of games, and computer technology; *for the engineer* who must have some understanding of the methods and terminology used in quality control, operations research, reliability work, experimental design, and computer technology; *for the research worker* who must know enough of the language of statistics to understand technical reports in his field; and above all *for the student of statistics on the elementary and intermediate level* who must become familiar with the language in which the methods, concepts, and ideas of modern statistics are expressed.

In compiling the basic lists of definitions and formulas, the authors went over literally hundreds of books at various levels of difficulty on both statistics and related subjects. In addition to the well-established terminology, methods, and formulas, to which one is introduced by most basic textbooks on statistics, the material chosen for inclusion in this book reflects some of the recent developments in the areas of decision theory, experimental design, the theory of games, computer techniques, operations research, and statistical inference.

This book consists of two parts: Part I is a dictionary of statistical terms and Part II is an outline of statistical formulas. One of the reasons why we did not incorporate the formulas among the definitions was to avoid repetitive explanations of the notation, repetitive formulations of basic assumptions, and repetitive references and other comments. All these things will be found at the beginning of the appropriate sections of Part II. *We suggest that, in order to use this book most efficiently, the reader should always look first for the information he seeks in Part I. There he will find, for over 1,000 terms, either a definition or a reference to another Part I entry. Where appropriate, entries in this part also include references to a formula, or formulas, in Part II; to a table; or to a book, or books, listed on pages 193, 194, and 195.*

The authors are very much indebted to their many friends and colleagues whose critical comments on various drafts and parts of the manuscript for this book have greatly contributed to its final form. The authors are also indebted to the Literary Executor of the late Sir Ronald A. Fisher, F.R.S., Cambridge, and to Oliver and Boyd, Edinburgh, for their permission to reprint the table of the t distribution; to Professor E. S. Pearson and the Biometrika trustees for permission to reproduce the tables of the chi-square and F distributions from their *Biometrika Tables for Statisticians;* and to the American Society for Testing and Materials for their permission to reproduce the table of control chart constants.

John E. Freund

Frank J. Williams

CONTENTS

GREEK ALPHABET

Α	α	alpha	Ν	ν	nu
Β	β	beta	Ξ	ξ	xi
Γ	γ	gamma	Ο	o	omicron
Δ	δ	delta	Π	π	pi
Ε	ϵ	epsilon	Ρ	ρ	rho
Ζ	ζ	zeta	Σ	σ	sigma
Η	η	eta	Τ	τ	tau
Θ	θ	theta	Υ	υ	upsilon
Ι	ι	iota	Φ	ϕ	phi
Κ	κ	kappa	Χ	χ	chi
Λ	λ	lambda	Ψ	ψ	psi
Μ	μ	mu	Ω	ω	omega

Part I
DICTIONARY OF BASIC STATISTICS

All references in boldface type are to formulas in Part II. For example, **(B.8d)** *is a reference to the computing formula for the standard deviation of grouped data on page 129.*

Part I
DICTIONARY OF BASIC STATISTICS

A

Absolute Deviation from the Mean – *See* Deviation from the Mean

Acceptable Quality Level, AQL – The quality (per cent defective) of a lot regarded as sufficiently good so that it is to be rejected only with a low probability, usually 0.05. The probability that a lot of AQL quality is rejected is called the "producer's risk."

Acceptance Number – (1) A term widely used in acceptance sampling, where the decision whether to accept or reject a lot depends on the number of defectives observed in a sample. In *single sampling*, a lot is accepted if the number of defectives in the sample is less than or equal to the *acceptance number* c, and it is rejected if the number of defectives is greater than or equal to the *rejection number* $c + 1$. In *multiple sampling*, a lot is accepted or rejected if the number of defectives in the first sample is less than or equal to the *acceptance number* c_1 or greater than or equal to the *rejection number* c_1'; if no decision can be reached on the basis of the first sample, the lot is accepted or rejected if the combined number of defectives in the first and second samples is less than or equal to the *acceptance number* c_2 or greater than or equal to the *rejection number* c_2'; if no decision can be reached on the basis of the first two samples, the procedure is continued with a third sample, perhaps a fourth sample, and so on. (2) In sequential sampling, a hypothesis is accepted after the kth observation if no decision was reached prior to that time and the total of the first k observations is less than or equal to (or greater than or equal to, depending on the hypothesis) the *acceptance number* a_k; correspondingly, the hypothesis is rejected after the kth observation if no decision was reached prior to that time and the total of the first

Acceptance Number—(continued) k observations is greater than or equal to (or less than or equal to, depending on the hypothesis) the *rejection number* r_k.

Acceptance Region—The subset of the sample space (or the set of all values of the test statistic) for which a null hypothesis is accepted.

Acceptance Sampling—The use of statistical sampling in the inspection and subsequent acceptance or rejection of raw materials, lots of manufactured products, etc. *See also* Acceptance Number; Multiple Sampling; Sequential Sampling.

Accuracy—The tendency of values of an estimator to come close to the quantity they are intended to estimate. *See also* Precision

Admissible Decision Function—A decision function is said to be admissible if there is no other decision function that is *uniformly better;* that is, if there is no other decision function which, according to some criterion, is sometimes better but never worse.

Admissible Strategy—A strategy which is not dominated by any other pure or randomized strategy; that is, a strategy is admissible if there exists no other strategy which, according to some criterion, is sometimes better but never worse.

Aggregative Index—*See* Simple Aggregative Index; Weighted Aggregative Index

ALGOL (ALGOrithmic Language)—A procedure-oriented, algebraic computer programming language designed for use in solving mathematical and engineering problems. Although a somewhat more powerful language than FORTRAN, ALGOL has not yet been widely adopted in the United States.

Algorithm—A well-defined procedure which, when routinely applied, leads to a solution of a particular class of mathematical problems. Various algorithms exist, for example, for solving linear programming problems.

Alias—*See* Confounding

Alienation, Coefficient of—A measure of the lack of linear association between two variables; it is given by the square root of 1 minus the square of the coefficient of correlation. **I.19.** Usually designated by k, its square is the coefficient of nondetermination.

Allocation of Sample—The assignment of portions of a sample to different parts of a population. *See also* Equal Allocation; Optimum Allocation; Proportional Allocation

Alpha, α—*See* Level of Significance; Regression Coefficient; Type I Error

Alpha-Four, α_4—*See* Kurtosis

Alpha-Three, α_3—*See* Skewness

Alternative Hypothesis—The hypothesis which one accepts when the

Alternative Hypothesis – (continued) null hypothesis (the hypothesis under test) is rejected. It is usually denoted H_A or H_1.

Analog Computer – Broadly speaking, an analog computer is any mathematical instrument which performs calculations not with numbers directly, but with physical quantities which represent (serve as "analogies" to) variables undergoing calculations. Examples of such physical quantities are lengths as in a slide rule, lengths and areas as in a planimeter, or ohms and volts as in electrical models. In contrast, a *digital computer* uses numbers, as such, to represent these variables.

Analysis of Covariance – A statistical analysis which consists of the combined application of linear regression and analysis of variance techniques. It is used when treatments are compared in the presence of concomitant variables which can be neither eliminated nor controlled. Further information may be found in the book by R. L. Wine listed on page 195.

Analysis of Variance, ANOVA – (1) The analysis of the total variability of a set of data (measured by their total sum of squares) into components which can be attributed to different sources of variation. A table which lists the various sources of variation together with the corresponding degrees of freedom, sums of squares, mean squares (sometimes also expected mean squares), and values of F, is called an *analysis of variance table.* (2) Sometimes abbreviated ANOVA, the term also refers to the totality of statistical techniques based on this kind of analysis. *See also* One-Way Analysis of Variance; Two-Way Analysis of Variance

Analysis of Variance Table – *See* Analysis of Variance

Annual Rate – A quantity designed to facilitate the reporting of month-to-month changes in data that are ordinarily (and more meaningfully) reported on an annual basis. An annual rate is calculated by multiplying a deseasonalized monthly figure by twelve. For instance, if a store's July sales were \$1.4 million and the July value of the corresponding seasonal index is 0.80, it can be said that these July sales were running at an annual rate of $\frac{1.4}{0.8} \cdot 12 = \$21$ million.

Annual Trend Increment – The average year-to-year change produced in a series of data by the action of forces (such as population growth or advances in technology) which presumably exert their influence over long periods of time.

A Posteriori Probability – A probability assigned to an event which takes into account specific sample data relating directly to the event; the term is usually applied to the probabilities $P(B_r|A)$ in Bayes' formula,

A Posteriori Probability – (continued)
D.11. A posteriori probabilities are also referred to as *posterior* or *inverse* probabilities; in the continuous case one speaks of *a posteriori* or *posterior densities.* **D.11b.**

Approximate – To approximate is to obtain a result near a desired result, or to obtain a succession of results approaching a desired result. An approximate result is one that is nearly but not exactly correct.

Approximation – A result that is not exact, but sufficiently close for a given purpose. For instance, 22/7 may serve as an approximation to the value of π. In statistics, the normal distribution may be used as a large-sample approximation of the binomial distribution.

A Priori Probability – A probability assigned to an event prior to the availability of specific sample data relating directly to the event; the term is usually applied to the probabilities $P(B_r)$ in Bayes' formula. **D.11.** A priori probabilities are also referred to as *prior* or *antecedent* probabilities; in the continuous case, one speaks of *a priori* or *prior densities.* **D.11b.**

Arc Sine Law – The limiting distribution of the proportion of the time one player leads in "Heads or Tails" as the number of tosses becomes infinite. A detailed discussion of this law may be found in the book by W. Feller listed on page 194.

Arc Sine Transformation – The transformation $y = \arcsin\sqrt{x}$ is often used to make data consisting of proportions (or frequencies) amenable to analysis of variance or regression techniques. Special tables for this transformation may be found in the book by G. W. Snedecor listed on page 195.

Area Sampling – A method of sampling where a geographical region is subdivided into smaller areas (counties, villages, city blocks, etc.), some of these areas are selected at random, and the chosen areas are then subsampled or surveyed 100 per cent.

Arithmetic Line Chart – A graph obtained by plotting values of a time series on arithmetic paper (graph paper with uniform subdivisions for both scales) and connecting successive points with straight lines.

Arithmetic-Logic Section (of a Digital Computer) – *See* Digital Computer

Arithmetic Mean – Commonly referred to as an "average," the arithmetic mean (or simply the mean) of n numbers is given by their sum divided by n. **A.1** and **B.1.** *See also* Mean

Arithmetic Mean of Price Relatives – An index number which is obtained by multiplying by 100 the arithmetic mean of the price relatives of a given set of commodities. **K.4.**

Arithmetic Paper – A graph paper with uniform subdivisions for both scales; on either scale, equal distances represent equal amounts.

Arithmetic Probability Paper—A graph paper, also called *normal probability paper*, with uniform subdivisions for one scale and the other scale ruled in such a way that a cumulative normal distribution plots as a straight line.

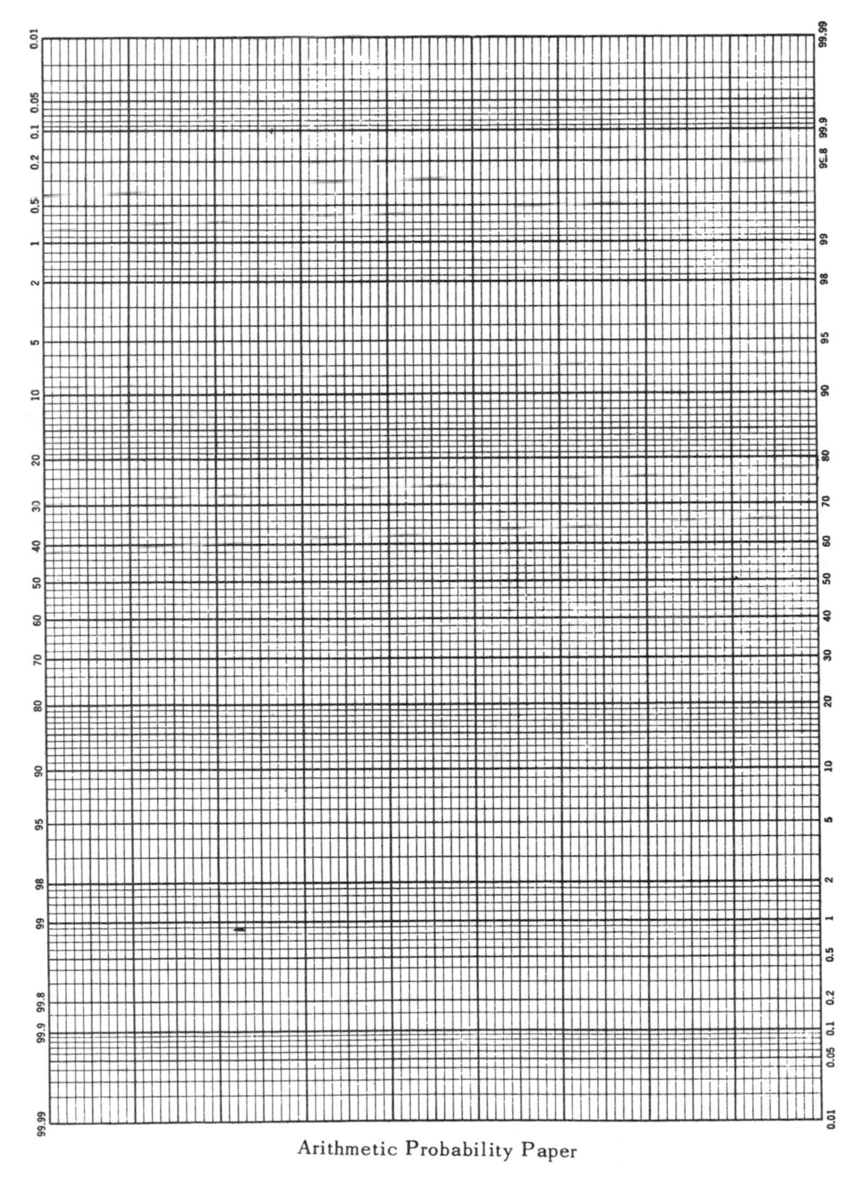

Arithmetic Probability Paper

Array – A particular arrangement of observations, say, according to size. *See also* Matrix

Arrival Distribution – In queuing theory, a probability function for the number of customers arriving at a service counter during a given period of time. The Poisson distribution is often used as a model for arrival distributions. *See also* Interarrival Distribution

Arrival Rate, Mean – In queuing theory, the average number of customers arriving at a service counter per unit time.

Assignable Variation – In quality control, variation from a standard that is not due to chance, but results from such detectable (hence, assignable) causes as substandard raw materials, faulty machine settings, poorly trained operators, and the like.

Assignment Problem – A particular kind of linear programming problem in which it is required to distribute n elements among n cells, one element per cell, in such a way that, given the cost (benefit) of each individual assignment, the total cost (benefit) realized from the complete assignment is minimized (maximized).

Association, Coefficient of – A measure of independence for data grouped in a two-by-two (contingency) table. **G.16.** Usually designated by Q, its limiting values are -1 and $+1$, corresponding to perfect negative and positive association.

Association Scheme – For partially balanced incomplete block designs, an arrangement which shows what treatments do and do not appear together in a block.

Assumed Mean – The origin of a scale used when coding to simplify the calculation of a mean, a standard deviation, or some other statistical description. It is designated x_0 in **B.1.**

Asymmetrical Distribution – *See* Skewness

Asymptotic Efficiency – The limiting efficiency of a sequence of estimators as the sample size becomes infinite.

Asymptotically Efficient – *See* Efficient

Attenuation, Correction for – In psychology and education, a correction sometimes applied to (raw) correlation coefficients which, due to errors of measurement, tend to be reduced or "attenuated." The correction provides a better estimate of the correlation which would exist between two traits if perfect (that is, errorless) measurements were available. For further information see the first book by A. L. Edwards listed on page 193.

Attributes, Inspection by – In quality control, inspection in which the quality of an item is determined by noting whether it does or does not possess a given qualitative characteristic or attribute; for example, by noting whether or not a bolt has a head.

Autocorrelation—The internal correlation (relationship) between elements of a stationary time series; it is usually expressed as a function of the time lag between observations. For further information see the book by M. S. Bartlett listed on page 193. *See also* Correlogram; Serial Correlation

Automatic Coding Systems—Certain computer softwear items which facilitate the use of a computer itself in preparing a computer program. Of the major forms of automatic coding, the *compiler* is the most powerful and versatile.

Average—A term lacking in precision and having different colloquial meanings (as in "batting average," "average" taste, etc.), but widely used to refer to the arithmetic mean.

Average Deviation—*See* Mean Deviation

Average Outgoing Quality, AOQ—The average fraction defective in submitted lots of product which are finally accepted, including those accepted by the sampling plan and those initially rejected but accepted after all defective units have been "screened out," that is, replaced by good units. An AOQ depends on the per cent of defectives in the incoming lots and the proportion of defectives which the inspection procedure actually removes.

Average Outgoing Quality Limit, AOQL—As the per cent of defectives in submitted lots of product increases from zero, the average outgoing fraction defective initially increases, reaches a maximum value, and then decreases as more and more defectives are "screened out." The maximum value of the AOQ's is called the Average Outgoing Quality Limit, or simply the AOQL.

Average Sample Number, ASN—In multiple and sequential sampling, the expected sample size; that is, the ASN is the average number of observations required to reach a decision concerning the acceptance or rejection of a lot. The average sample number depends on the quality of the lot, and if one plots it for various values of the fraction defective, one obtains an *ASN-curve*.

Axioms of Probability—*See* Probability, Postulates of

B

Balanced Design—The balance of an experimental design depends on the allocation of treatments (levels of variables or factors) to blocks. A randomized block design is balanced if every treatment appears in each block the same number of times; an incomplete block design is balanced if every two treatments appear together in a block the same number of times. For further information see the book by W. G. Cochran and G. M. Cox listed on page 193.

Bar Chart—A chart, used to present frequency distributions or time series, which consists of bars (rectangles) of equal width, whose lengths are proportional to the frequencies (or values) they represent.

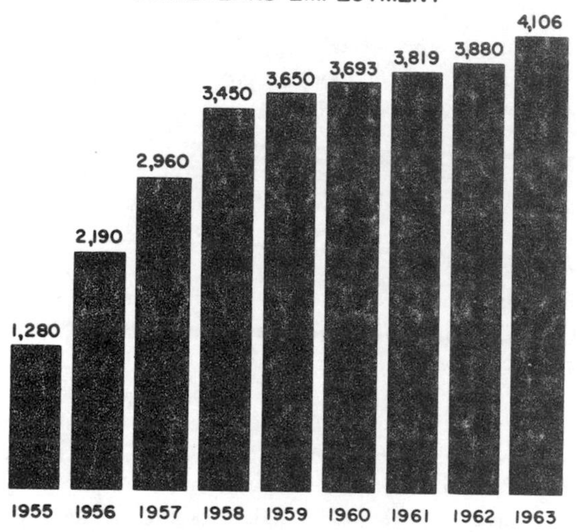

Bar Chart

Bartlett's Test—A test for the homogeneity of variances; specifically, a test of the null hypothesis that a set of independent random samples comes from normal populations having the same variance. A description of this test may be found in the book by E. S. Keeping listed on page 194.

Base Year (or Base Period)—In index number construction, the year (or period) with reference to which a comparison is being made. It is usually denoted by the subscript "o;" for example, the price of a commodity in the base year is written p_0 and the corresponding quantity produced (consumed, sold, etc.) is written q_0.

Bayes' Formula—A formula for the calculation of certain conditional probabilities sometimes referred to as probabilities of "causes." D.11. It enables one to express a posteriori probabilities in terms of a priori probabilities and other known probabilities, and, hence, it enables one to revise probabilities in the light of new evidence. The formula plays a fundamental role in some approaches to statistical inference.

Bayes Risk—*See* Decision Theory

Bayes Solution—*See* Decision Theory

Bayesian Estimate—An estimate based on the a posteriori probabilities of a parameter obtained with the use of Bayes' formula. For certain loss functions, the mean and the mode of such a posteriori distributions have been used as Bayesian estimates of a parameter.

Bayesian Inference – A form of inference utilizing Bayes' formula, **D.11**, and generalizations. It is distinguished by the fact that parameters, looked upon as random variables, are assigned a priori distributions. Detailed discussions of this kind of inference may be found in the books by D. V. Lindley, L. J. Savage, and R. Schlaifer listed on pages 194 and 195.

Bayesian Statistics – Statistical methods which utilize prior information (that is, objective or subjective collateral information) about parameters. The term has also been applied by some to all statistical methods based on the concept of subjective, or personal, probability.

Behrens-Fisher Problem – The problem of testing for the equality of the means of two normal populations having unequal variances; it is named after a solution involving fiducial probabilities proposed by Behrens and advocated by Fisher. **F.21**,

Bell-Shaped Distribution – A distribution having the over-all shape of a vertical cross section of a bell; normal distributions are among those having this characteristic.

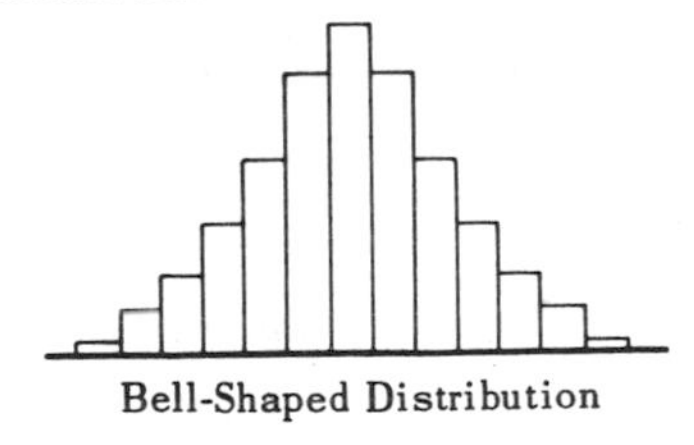

Bell-Shaped Distribution

Bernoulli Random Variable – A random variable whose values, 0 and 1, correspond to "failure" and "success" in a Bernoulli trial.

Bernoulli Trial – A mathematical model for an experiment having just two outcomes, usually referred to as "success" and "failure." It is a special case of the binomial distribution, **E.23**, with $n = 1$.

Best Estimate – This term is sometimes applied to minimum-variance unbiased estimates. *See also* Consistency; Minimum-Variance Estimator; Power Efficiency; Relative Efficiency; Sufficiency; Unbiased Estimator

Best Fit – *See* Goodness of Fit; Least Squares, Method of

Beta, β – *See* Regression Coefficient; Type II Error

Beta Distribution – A distribution which is closely related to the F distribution, used extensively in analysis of variance. **E.30**. In Bayesian inference, the beta distribution is sometimes used as the a priori distribution for the parameter θ, the probability of a success on an individual trial, of the binomial distribution.

Between-Samples Sum of Squares – The treatment sum of squares in a one-way analysis of variance, (**J.1d**); measuring the variability among

Between-Samples Sum of Squares – (continued) the sample means, it serves as an indication of possible differences among the corresponding population means.

Bias – (1) In problems of estimation, an estimator is said to be biased if its expected value does *not* equal the parameter it is intended to estimate. (2) In index number construction, the bias of an index is the systematic tendency to overestimate or underestimate changes. (3) In sampling, a bias is a systematic error introduced by selecting items from a wrong population, favoring some of the elements of a population, or poorly phrasing questions. (4) In hypothesis testing, a test is said to be biased when the probability of rejecting the null hypothesis is *not* a minimum when the null hypothesis is, in fact, true.

Bimodal Distribution – *See* Multimodal Distribution

Binary System – A number system in which numbers are written to the base 2. For example, in the binary system 11011 represents the number $1 \cdot 2^4 + 1 \cdot 2^3 + 0 \cdot 2^2 + 1 \cdot 2^1 + 1 \cdot 2^0 = 27$, and 0.1101 represents the number $0 \cdot 2^0 + 1 \cdot 2^{-1} + 1 \cdot 2^{-2} + 0 \cdot 2^{-3} + 1 \cdot 2^{-4} = 1/2 + 1/4 + 1/16 = 13/16$. The binary system is used extensively in connection with digital computers.

Binomial Coefficient – The binomial coefficient $\binom{n}{k}$ is the coefficient of $a^k b^{n-k}$ in the expansion of $(a + b)^n$; in factorial notation it is given by $\frac{n!}{k!(n-k)!}$. The calculation of binomial coefficients can be facilitated by using the following arrangement, called *Pascal's triangle:*

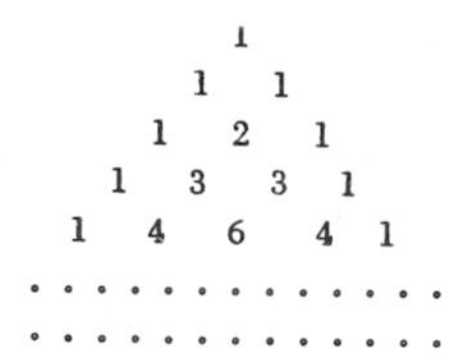

Each row begins and ends with a 1, each other number is the sum of the two nearest numbers in the row immediately above, and the binomial coefficient $\binom{n}{k}$ is given by the $(k + 1)$ st number in the $(n + 1)$ st row.

Binomial Distribution – The distribution of the number of *successes* in n *trials*, when the probability of a success remains constant from trial to trial and the trials are independent. E.23. Also, the distribution of the sum of n independent Bernoulli random variables, each with the same probabilities θ and $1 - \theta$ for success and failure. Values of the

Binomial Distribution—(continued)
binomial probability function may be obtained from the National Bureau of Standards Tables and those by H. G. Romig listed on page 195.

Binomial Index of Dispersion—A chi-square statistic used to test whether k samples come from populations having binomial distributions with the same parameter θ. G.11.

Binomial Probability Paper—A special kind of graph paper used in the analysis of count (or enumeration) data. The units are marked so that along the axes distances from the origin are proportional to the square roots of the coordinates. The use of this kind of graph paper is discussed in the book by W. J. Dixon and F. J. Massey listed on page 193.

Bio-Assay—Methods devoted to the design and analysis of tests made for the purpose of assigning limits within which the potency of such preparations as vitamins, drugs, and sera may be presumed to fall in comparison with a standard. A detailed discussion of this subject may be found in the book by C. W. Emmens listed on page 193. *See also* Logit; Probit

Birth and Death Process—A random process representing the size of a population, for which in a very small time interval the population size increases by one, decreases by one, or stays the same. It applies, for example, to populations of cells which can split or die.

Birth Process, Pure—*See* Pure Birth Process

Biserial Correlation Coefficient—A measure of the strength of the relationship between two variables, one continuous and the other recorded as a dichotomy. I.28. It is assumed that the dichotomized variable is really continuous and normally distributed. *See also* Point Biserial Coefficient of Correlation

Bit—(1) A common abbreviation for "Binary Digit," that is, 0 or 1 in the binary system. (2) In information theory, when a message consists of some pattern of the digits 0 and 1, each digit conveys a "unit of information," and each unit of information is called a "bit." Hence, the term "bit" refers both to a two-valued variable used to represent a unit of information, and to a unit of information itself.

Bivariate Distribution—*See* Joint Density Function; Joint Distribution; Joint Distribution Function; Joint Probability Function

Bivariate Normal Distribution—The joint distribution of two random variables X and Y for which the marginal distribution of X is normal, the regression of Y on X is linear, and the conditional distribution of Y for a fixed value of X is normal with a variance which does not depend on the value of X. E.41. If X and Y are furthermore independent and have the same variance, their joint distribution is referred to as the *circular normal distribution;* it owes its name to the fact that

Bivariate Normal Distribution—(continued) contours of equal probability density are, in fact, circles (while in the general case they are ellipses).

Block—A homogeneous grouping of experimental units, designed to enable the experimenter to isolate (hence, eliminate) variability due to extraneous causes.

Block Effect—In analysis of variance, a quantity (usually a parameter of the model), which represents the change in response produced by a given block. In (J.3a) the block effects are the parameters β_j.

Block Sum of Squares—In analysis of variance, that component of the total sum of squares which can be attributed to possible differences among the blocks. (J.3e).

Borel Field—*See* σ-Field

Branching Process—Also called a *Cascade, Multiplicative,* or *Chain-reaction process,* a branching process represents a population for which from generation to generation each individual gives rise to 0, 1, 2, 3, . . . new individuals. It applies, for example, to genetics, where a gene has a chance to reappear in 0, 1, 2, 3, . . . direct descendants of an organism.

Brownian Motion—The irregular, continuous motion or movement of a microscopic particle immersed in a fluid or gas, which results from the bombardment of the particles by the molecules of the liquid or gas. The term is also used for random processes which serve as models for this kind of motion.

Bunch Maps—A set of charts used in econometrics as the basis of a graphical technique to determine whether or not multicollinearity is present in a multiple correlation or regression analysis of economic data.

BLS Seasonal Factor Method (1964)—An iterative procedure, prepared by the Bureau of Labor Statistics, for developing seasonal factors (a seasonal index) for its employment and unemployment series. The method has yielded good results when applied to various other general economic series, and the program (prepared for processing on an IBM 1401-1410 tape system) is available from the Bureau at a modest charge.

Business Cycle—A business cycle consists of a repeated up-and-down movement of business activity, having a length greater than one year. The periods of prosperity, recession, depression, and recovery, which constitute the four phases of a complete cycle, are considered to be caused by much more complex factors than the weather, social customs, etc., which account for seasonal variations.

C

c **Chart**—In quality control, a control chart for the number of defects observed per unit (which might be a single item, part of an item, or a group of items). L.13 and L.14.

Cascade Process—*See* Branching Process

Categorical Distribution—A frequency distribution in which items are grouped into nonoverlapping categories according to some qualitative description. For instance, when classifying the books published in a given year, one might construct a categorical distribution showing how many were historical novels, how many were mysteries, how many were cookbooks, and so on.

Categorical Scaling—*See* Scaling.

Cauchy Distribution—Although the graph of this distribution has the general bell-shaped appearance of a normal curve, it has the distinguishing feature that its moments do not exist. E.31.

Cell Frequency—In the analysis of count (or enumeration) data, the number of items falling into an individual category (classification, or subclassification) is sometimes referred to as a cell frequency. The term is used mainly in connection with contingency tables.

Censored Data—This term is used when only the k smallest values, the k largest values, or the remaining values of a sample of size n are actually recorded, where k is specified in advance. If values exceeding a fixed constant and/or less than a fixed constant are *not* recorded, the data are referred to as *truncated.*

Census—A complete enumeration of a population; partial enumerations based on samples are sometimes referred to as "sample censuses."

Centered (12-Month) Moving Average—A moving average obtained by averaging successive pairs of values of a 12-month moving average; the purpose of this adjustment is to obtain values corresponding to the midpoints (rather than the beginning or the end) of the various months.

Central Limit Theorem—In its simplest form, the theorem states that for random variables from a population with a finite variance, the sampling distribution of the standardized sample mean approaches the standard normal distribution as the sample size n becomes infinite. This theorem is of fundamental importance in probability and statistics, as it justifies the application of normal distribution theory to a great variety of statistical problems.

Central Line—*See* Control Chart

Central Tendencies, Measures of—A name sometimes given to "averages," that is, to statistics such as the mean, median, or mode.

Chain Index – An index number in which figures for any one period are compared with those of the preceding period; such individual comparisons are often "chained" back to a fixed base period by a process of successive multiplication, hence, the name "chain index."

Chain-Reaction Process – *See* Branching Process

Chance Variable – *See* Random Variable

Chance Variation – In general, the term applies to the fluctuations one observes between the values of a random variable. An appreciation of the fact that such fluctuations can be studied mathematically and the fact that one can make predictions about their possible size is basic to an understanding of modern statistics.

Change of Variable (or Scale) – *See* Coding; Transformation

Chapman-Kolmogorov Equation – A fundamental relation satisfied by the transition probabilities of a Markov chain. For detailed information, see the book by W. Feller listed on page 194.

Characteristic Function – In probability theory, a function associated with a random variable X, whose values $\varphi(t)$ equal the expected value of e^{itX}; the coefficient of $(it)^k/k!$ in its series expansion equals the kth moment of X about the origin. Observe that in matrix theory and in measure theory the term "characteristic function" has an entirely different meaning. E.10.

Chebyshev's Theorem (or Inequality) – An important result in probability, which provides an upper limit to the probability that a value of a random variable differs from the mean of its distribution by more than k standard deviations. If $P(|X - \mu| > k\sigma)$ denotes the probability that the value assumed by a random variable X differs from its mean μ by more than $k\sigma$, the theorem can be expressed symbolically in the following form:

$$P(|X - \mu| > k\sigma) < \frac{1}{k^2}$$

When applied to actual data, Chebyshev's Theorem correspondingly provides an upper limit to the *proportion* of the data which differs from the mean by more than k standard deviations.

Chi-Square Distribution (χ^2 Distribution) – A distribution which is of great importance in inferences concerning population variances or standard deviations. E.32. It arises in connection with the sampling distribution of the sample variance for random samples from normal populations; it is a special form of a *gamma distribution*. The parameter of this distribution is ν, the number of degrees of freedom. The table on page 15 contains values of $\chi^2_{\alpha,\nu}$, which denotes the value for which the area *to its right* under the chi-square distribution with ν degrees of freedom is equal to α.

The Chi-Square Distribution*

The entries in this table are values of $\chi^2_{\alpha,\nu}$, for which the area to their right under the chi-square distribution with ν degrees of freedom is equal to α.

ν	$\alpha = .995$	$\alpha = .99$	$\alpha = .975$	$\alpha = .95$	$\alpha = .05$	$\alpha = .025$	$\alpha = .01$	$\alpha = .005$	ν
1	.0000393	.000157	.000982	.00393	3.841	5.024	6.635	7.879	1
2	.0100	.0201	.0506	.103	5.991	7.378	9.210	10.597	2
3	.0717	.115	.216	.352	7.815	9.348	11.345	12.838	3
4	.207	.297	.484	.711	9.488	11.143	13.277	14.860	4
5	.412	.544	.831	1.145	11.070	12.832	15.086	16.750	5
6	.676	.872	1.237	1.635	12.592	14.449	16.812	18.548	6
7	.989	1.239	1.690	2.167	14.067	16.013	18.475	20.278	7
8	1.344	1.646	2.180	2.733	15.507	17.535	20.090	21.955	8
9	1.735	2.088	2.700	3.325	16.919	19.023	21.666	23.589	9
10	2.156	2.558	3.247	3.940	18.307	20.483	23.209	25.188	10
11	2.603	3.053	3.816	4.575	19.675	21.920	24.725	26.757	11
12	3.074	3.571	4.404	5.226	21.026	23.337	26.217	28.300	12
13	3.565	4.107	5.009	5.892	22.362	24.736	27.688	29.819	13
14	4.075	4.660	5.629	6.571	23.685	26.119	29.141	31.319	14
15	4.601	5.229	6.262	7.261	24.996	27.488	30.578	32.801	15
16	5.142	5.812	6.908	7.962	26.296	28.845	32.000	34.267	16
17	5.697	6.408	7.564	8.672	27.587	30.191	33.409	35.718	17
18	6.265	7.015	8.231	9.390	28.869	31.526	34.805	37.156	18
19	6.844	7.633	8.907	10.117	30.144	32.852	36.191	38.582	19
20	7.434	8.260	9.591	10.851	31.410	34.170	37.566	39.997	20
21	8.034	8.897	10.283	11.591	32.671	35.479	38.932	41.401	21
22	8.643	9.542	10.982	12.338	33.924	36.781	40.289	42.796	22
23	9.260	10.196	11.689	13.091	35.172	38.076	41.638	44.181	23
24	9.886	10.856	12.401	13.848	36.415	39.364	42.980	45.558	24
25	10.520	11.524	13.120	14.611	37.652	40.646	44.314	46.928	25
26	11.160	12.198	13.844	15.379	38.885	41.923	45.642	48.290	26
27	11.808	12.879	14.573	16.151	40.113	43.194	46.963	49.645	27
28	12.461	13.565	15.308	16.928	41.337	44.461	48.278	50.993	28
29	13.121	14.256	16.047	17.708	42.557	45.722	49.588	52.336	29
30	13.787	14.953	16.791	18.493	43.773	46.979	50.892	53.672	30

*Based on Table 8 of *Biometrika Tables for Statisticians, Vol. I*, by permission of the *Biometrika* trustees.

Chi-Square Statistic – A statistic which is given by a sum of terms, where each term is the quotient of the squared difference between an observed frequency and an expected frequency divided by the expected frequency. *See also* Binomial Index of Dispersion; Contingency Table; Goodness of Fit; Poisson Index of Dispersion

Circular Normal Distribution – *See* Bivariate Normal Distribution

Class Boundary – The dividing line between successive classes of a frequency distribution; to avoid ambiguities, class boundaries are usually chosen so that they represent "impossible" values, namely, values which cannot occur among the data which are to be grouped. Class boundaries are also referred to as *real class limits*. *See also* Frequency Distribution

Class Frequency – The number of items falling into a particular class of a frequency distribution. *See also* Frequency Distribution

Class Interval – The length of a class, or the range of values covered by a class of a frequency distribution. For any one class, it is given by the difference between its upper and lower boundaries; for a distribution with equal class intervals it is given by the difference between successive class marks or class boundaries. *See also* Frequency Distribution

Class Limit – The upper and lower limits of a class are, respectively, the largest and smallest values it can contain. *See also* Frequency Distribution

Class Mark – The midpoint of a class; hence, the mean of its boundaries or the mean of its limits. *See also* Frequency Distribution

Classical Statistical Inference – *See* Statistical Inference

Closeness of an Estimate – *See* Best Estimate; Consistency; Minimum-Variance Estimator; Power Efficiency; Relative Efficiency; Sufficiency; Unbiased Estimator

Cluster Sampling – A method of sampling in which the elements of a population are arranged in groups (or clusters); some of the clusters are selected at random, and the ones chosen are then subsampled, or surveyed 100 per cent. Generally, the clusters consist of natural groupings, and if they are geographic regions the sampling is referred to as *area sampling.*

COBOL (COmmon Business Oriented Language) – An elaborate computer programming language, independent of any make or model of computer, which is oriented toward the solution of a wide range of business data processing problems such as, for example, the maintenance of large files of data. The COBOL system consists of a *source program* written in the COBOL language, a *compiler* to translate a *source program* into an *object program,* and a COBOL *library.*

Cochran's Test – A nonparametric test applicable to the analysis of three or more *matched* sets of frequencies or proportions. It applies, for example, to the analysis of an experiment where each of n individuals receives passing or failing grades for k different tasks. A discussion of this test may be found in the book by S. Siegel listed on page 195.

Cochran's Theorem – A theorem on quadratic forms which is of basic importance in the theory underlying the analysis of variance. A formal statement of the theorem may be found in the book by S. S. Wilks listed on page 195.

Coding – Transforming or changing a scale of measurement; that is, converting numbers given in one scale to their equivalents in a different scale. For instance, one might code the numbers 2.12, 2.15, 2.20, 2.10, and 2.18 by subtracting 2.10 from each and multiplying the differences by 100, thus getting the numbers 2, 5, 10, 0, and 8. Usually, the purpose of coding is to simplify the numbers with which one has to work, or to facilitate the comparison of data (distributions) given in different units of measurement. See, for example, (**B.1b**) and (**B.8d**). *See also* Normalized Standard Scores; Standard Units

Coding (a Digital Computer) – The act of translating a sequence of well-defined operations (which specify the particular way in which a problem is to be solved) into a set of detailed instructions appropriate for a given computer. Computer coding is often called *programming* and the resulting set of instructions is called a *computer code, computer program,* or *routine.*

Coding System – In information theory, any consistent scheme used to represent a given set of data. A coding system is usually employed either to reduce error or increase efficiency in the transmission of information.

Coefficient – In algebra, the term "coefficient" is most commonly used for constant factors as distinguished from variables; it is in this sense that one speaks of a regression coefficient. The term is also used to denote a dimensionless description of a set of data, or a distribution; for example, a coefficient of correlation or a coefficient of variation. *See also* individual listings under subject matter; for example, for Coefficient of Variation *see* Variation, Coefficient of

Combination – A selection of one or more of a set of distinct objects without regard to order. The number of possible combinations, each containing r objects, that can be formed from a collection of n distinct objects is given by $\frac{n!}{(n-r)!r!}$ and it is denoted $\binom{n}{r}$, ${}_nC_r$, C^n_r, or $C(n,r)$.

Combination – (continued)
For instance, the number of combinations of the first ten letters of the alphabet, taken three at a time, is $\frac{10!}{7!3!} = 120$. *See also* Binomial Coefficient.

Compiler – Generally the most useful type of automatic coding system, a compiler is a computer program which translates and expands (by adding required subroutines, for example) a source program into an object program. The FORTRAN and COBOL compilers are well-known automatic compiling systems in the scientific and business data processing areas, respectively. FORTRAN, COBOL, and some other languages, are often called "compiler languages."

Complement (of a Set) – The complement of a set A relative to a sample space S, denoted A' or $\bar{A}$, is the set which consists of all elements of S that do not belong to A.

Complete Block Design – An experimental design in which every treatment appears the same number of times in each block.

Complete Class of Decision Functions – In decision theory, a class of decision functions is said to be complete if for every decision function *not* in the class there is one in the class which is uniformly better (according to some criterion). A complete class of decision functions is referred to as *minimal* if all of the decision functions it contains are admissible. *See also* Admissible Decision Function; Uniformly Better Decision Function

Completely Randomized Design – An experimental design in which the treatments are allocated to the experimental units (plots) entirely at random.

Component Bar Chart – A bar chart in which the individual bars (rectangles) are divided into sections proportional in size to the components of the total they represent. The various components are usually shaded or colored differently to increase the over-all effectiveness of the chart. Such a chart is shown on page 19.

Components of a (Digital) Computer – *See* Digital Computer

Components of a Time Series – In time series analysis, the four basic types of movements: secular trend, seasonal variation, cyclical variation, and irregular variation. The combined effect of these components is presumed to account for the observed fluctuations and the over-all movement of a series.

Components of Variance Model – *See* Random-Effects Model

Composite Hypothesis – If the set of values a parameter can assume under a given hypothesis consists of more than one value, the hypothesis is said to be composite. Thus, if μ is the mean of a normal

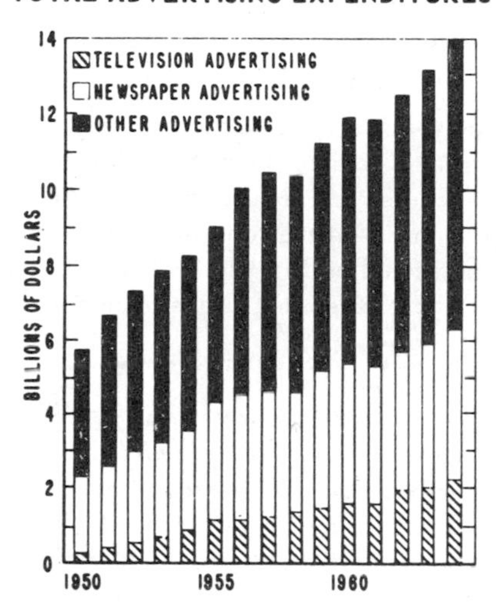

Component Bar Chart

Composite Hypothesis – (continued)

population with known variance, the hypothesis $\mu > \mu_0$ is composite, whereas the hypothesis $\mu = \mu_0$ is not. *See also* Simple Hypothesis

Compound Distribution – A distribution which arises when one observes a random mixture of random variables having the same general type of distribution; say, random variables having Poisson distributions with different values of the parameter λ.

Computer – *See* Analog Computer; Digital Computer

Computer Graphics – A combination of various communication and graphic arts skills, computer equipment, and computer techniques which results in the rapid and economical production of detailed drawings by a computer of such things as weather maps, cockpit and maneuver simulations, and different kinds of mathematical descriptions (for instance, parts described by means of arithmetic and two-dimensional geometric statements).

Computer Operating System – *See* System, Computer Operating

Concordance, Coefficient of – A measure of the agreement among k rankings of n individuals or items. **I.31.** The coefficient of concordance is linearly related to the mean of the pairwise rank correlation coefficients.

Conditional Distribution – The distribution of a random variable (or the joint distribution of several random variables) when the

Conditional Distribution – (continued) values of one or more other random variables are held fixed, or some other event has occurred. **E.19.**

Conditional Probability – If A and B are any two events and the probability of B is not equal to zero, then the conditional probability of A relative to B is denoted $P(A|B)$ and given by $P(A \cap B)/P(B)$, where $P(A \cap B)$ is the probability of the joint occurrence of A and B. **D.2.** Informally, the conditional probability of A relative to B is the probability that A will occur *given* that B has occurred or will occur.

Confidence Coefficient – *See* Confidence Interval

Confidence Interval – An interval for which one can assert with a given probability $1 - \alpha$, called the *degree of confidence* or the *confidence coefficient,* that it will contain the parameter it is intended to estimate. **F.10 - F.16, G.5 - G.7, I.10 - I.12,** and **I.24.** The endpoints of a confidence interval are referred to as the (*upper and lower*) *confidence limits;* they are generally values of random variables calculated on the basis of sample data. A confidence interval is said to be *one-sided* when only one of the limits is a value of a random variable, while the other limit is a constant (and often omitted) or infinite. For instance, a confidence interval for the proportion θ of defectives in a large lot may be given in the form $\theta < k$, where k is a value obtained from sample data; although it is not given explicitly, the lower limit of this confidence interval is 0.

Confidence Limits – *See* Confidence Interval

Confidence Set (or Region) – A generalization of the concept of a confidence interval, which applies to the simultaneous estimation of several parameters. See, for example, the discussion of confidence regions for the simultaneous estimation of the mean and the variance of a normal population in the book by S. S. Wilks listed on page 195.

Confounding – In factorial experimentation, a process by which one foregoes some information (usually about higher-order interactions) in order to reduce an experiment to manageable size. Specifically, two effects are confounded and referred to as *aliases,* if it is impossible to differentiate between them on the basis of a given experiment; that is, the contrasts which measure one effect are the very same contrasts which measure the other effect. An effect is confounded with blocks if it is impossible to differentiate between the effect and variations caused by differences among the blocks.

Consistency – An estimator (depending on the sample size n) is said to be consistent if the probability that it will assume a value arbitrarily close to the parameter it is intended to estimate *approaches one* when n becomes infinite. For instance, an unbiased estimator is consistent if the variance of its sampling distribution *approaches zero* when n becomes infinite.

Consumer Price Index – Constructed by the Bureau of Labor Statistics, this index is designed to measure the over-all change in the prices of a large collection of goods and services (called a "market basket"), which represents the greater part of the expenditures of city wage-earner and clerical-worker families. The primary source of this important index is the *Monthly Labor Review.*

Consumer's Risk – The probability β of committing a Type II error; the term is used especially in problems of sampling inspection.

Contagious Distribution – If the distribution of a random variable X depends on a parameter which is actually a value of a random variable, then the *unconditional* distribution of X is referred to as a contagious distribution. For instance, the distribution of the number of persons hurt in an automobile accident may be obtained as a contagious distribution from the distributions of the number of persons hurt in one-car accidents, in two-car accidents, in three-car accidents, . . . , and a distribution for the number of cars involved in an accident.

Contingency Coefficient – A measure of the strength of the association between two variables, usually qualitative, on the basis of data tallied into a contingency table. **G.14.** The value of this statistic is never negative and it has a maximum less than 1, depending on the number of rows and columns in the contingency table.

Contingency Table – A table consisting of two or more rows and two or more columns, into which individuals or items are classified according to two criteria (or variables). The simplest form of contingency table, the 2 by 2 table, arises when both variables are dichotomized. The notation used in connection with contingency tables and formulas for their analysis are given in **G.13.** The following is a 3 by 2 contingency table in which 237 spending units have been classified according to whether (1) their annual income was $5,000 or less, over $5,000 but less than $10,000, or $10,000 or more, and (2) they expected to be better off or worse off financially one year later:

Contingency Table – (continued)

	Better off	*Worse off*	
$5,000 or less	48 (59.2)	42 (30.8)	90
Over $5,000 but less than $10,000	63 (57.3)	24 (29.7)	87
$10,000 or more	45 (39.5)	15 (20.5)	60
	156	81	

The numbers 48, 42, 63, 24, 45, and 15 are the *observed (cell) frequencies*, the numbers in parentheses are the *expected (cell) frequencies* calculated according to (**G.13b**), and the numbers at the bottom and side of the table are called the *marginal totals*.

Continuity Correction – An adjustment made when approximating the distribution of a discrete random variable with that of a continuous random variable. For instance, when approximating a binomial distribution with a normal distribution, the adjustment consists of representing each integer k (from 0 to n) by a corresponding interval from $k - 1/2$ to $k + 1/2$. **G.8.**

Continuous Density – *See* Probability Density Function

Continuous Distribution – The distribution of a continuous random variable. *See also* Probability Density Function

Continuous Game – In game theory, a game where each player has a continuum of pure strategies; for example, when each player's strategy consists of choosing a real number between 0 and 1.

Continuous Population – It is customary to say that one is sampling from a continuous population when one observes values of a continuous random variable.

Continuous Random Variable – A random variable whose range (set of possible values) is an interval or a set of intervals on the real axis (often the entire real axis), and which possesses a probability density function. For instance, a random variable having a normal distribution is said to be continuous.

Continuous Sampling Inspection – A sampling procedure in which current inspection results determine whether inspection of the next items will be by sample or by 100 per cent inspection.

Contrast—In analysis of variance, a linear combination of observations which is designed to estimate a particular parameter of the model; the coefficients of the linear combination are subject to the restriction that their sum must equal zero. Two such contrasts are said to be *orthogonal* if the sum of the pairwise products of their coefficients is equal to zero.

Control Chart—In quality control, a chart used to decide periodically whether a process is in statistical control. The use of such a chart facilitates the detection and elimination of assignable causes of process variation, thereby reducing rejects and rework, improving product quality, and lowering inspection cost. The *central line* of a control chart corresponds to the mean of the sampling distribution of the statistic on the basis of which control is to be maintained, and it may be based on specifications or on an analysis of past data. L.2-L.14. *Control limits* are horizontal lines drawn on a control chart at appropriate distances above and below the central line; they are referred to, respectively, as the *upper control limit*, UCL, and the *lower control limit*, LCL. So long as plotted points fall between the control limits the process is presumed to be "in control;" a point which falls outside the control limits is presumed to indicate the presence of assignable causes of variation. When the control limits are set at three standard deviations above and below the central line, they are referred to as *three-sigma control limits;* for normal distributions, the probability of exceeding such limits by chance is 0.0026. Experience gathered from a wide variety of applications suggests the suitability of such limits as a criterion for action, namely, as a criterion for detecting assignable causes of variation.

Control Limits—*See* Control Chart

Control Section (of a Digital Computer)—*See* Digital Computer

Control System—*See* System, Control

Conventional Information Systems—A term often applied to present-day information systems based on ideas (suggested by A. M. Turing and J. von Neumann) of executable instructions, a computer in which instructions can be stored, and a program consisting of such instructions. Such systems, restricted to executing fixed commands (instructions), can handle all information processes (statistical data reduction, commercial data processing, etc.) in which the structure of the data is completely determined before processing and the necessary processing operations can be expressed in a computing procedure (algorithm) in final form.

Convolution—An integral (or sum) used to obtain the distribution of the sum of two or more random variables. A discussion of convolutions may be found in the book by J. E. Freund listed on page 194.

Correlated Samples – Two samples consisting of paired data, such as the ages of husbands and wives, or the weights of individuals before and after a diet.

Correlation – In general, the term denotes the relationship (association or dependence) between two or more qualitative or quantitative variables. *See also* Curvilinear Correlation; Linear Correlation; Multiple Correlation Coefficient; Negative Correlation; Partial Correlation; Positive Correlation; Rank Correlation

Correlation Analysis – The analysis of paired data constituting the values of two random variables X and Y; it is referred to as *normal correlation analysis* if X and Y have the bivariate normal distribution. More generally, the term applies to the analysis of n-tuples of data constituting the values of n random variables. *See also* Regression Analysis

Correlation Coefficient – (1) For two random variables, the ratio of their covariance to the product of their standard deviations; it is designated ρ (*rho*). E.22. (2) A measure of the linear relationship between two quantitative variables, known also as the *Pearson product-moment coefficient of correlation.* It is denoted by the letter r and its values range from -1 to $+1$, where 0 indicates the absence of any linear relationship, while -1 and $+1$ indicate, respectively, a perfect *negative* (inverse) and a perfect *positive* (direct) relationship. I.17. *See also* Alienation, Coefficient of; Determination, Coefficient of; Negative Correlation; Nondetermination, Coefficient of; Positive Correlation; z-Transformation

Correlation Matrix – The matrix whose elements are correlation coefficients; that is, for $i \neq j$ the element a_{ij} of the matrix is the correlation coefficient for the ith and jth variables, while $a_{ii} = 1$ for all i.

Correlation Ratio – A measure of the curvilinear (nonlinear) relationship between two variables. It is discussed in the book by M. G. Kendall and A. Stuart listed on page 194.

Correlation Table – A two-way frequency distribution; that is, a table designed to group, or classify paired quantitative data. Correlation tables are usually constructed to facilitate the calculation of correlation coefficients.

Correlogram – A graph of the serial correlation within a time series plotted as a function of the time lag between observations.

Cost Function – (1) In sampling theory, a function giving the cost of obtaining the data in terms of such factors as the number of observations made (perhaps, in various stages of a sampling plan), the number of units damaged or destroyed, overhead, etc. (2) In decision theory, a cost function constitutes one of the components of a loss function.

Count Data – Data obtained by performing actual counts, as constrasted to data obtained by performing measurements on continuous scales. Such data are also referred to as *enumeration data.*

Covariance – The expected value of the product of the deviations of two random variables from their respective means. E.21. The covariance of two *independent* random variables is zero, but a zero covariance does not imply independence. The covariance is also referred to as the *first product-moment,* and it is analogously defined for a set of paired data as the mean of the products obtained by pairwise multiplying the deviations from the respective means. As such, it appears in the numerator of the formula for the coefficient of correlation, and it is thus indicative of the extent of the linear association between the two variables.

Covariance Analysis – *See* Analysis of Covariance

Cramer-Rao Inequality – An inequality which provides a lower bound to the variance of the sampling distribution of an unbiased statistic. A precise formulation of this theorem as well as generalizations may be found in the book by E. S. Keeping listed on page 194.

Critical Region – For a given test, the subset of the sample space which contains all outcomes for which the null hypothesis is rejected. The *size* of a critical region is the probability of obtaining an outcome belonging to the critical region when the null hypothesis is true; hence, it is the probability α of a Type I error.

Critical Values – The dividing lines of a test criterion or, more generally, the boundary of a critical region.

Cross Correlation – *See* Lag Correlation

Cross-Section Data – Data collected on some (usually economic) variable, at the same point or period in time, from different geographical regions, organizations, institutions, etc; for instance, the grape tonnages harvested in October, 1965, in each of the forty-eight vineyard counties of California. *See also* Time Series

Cross Stratification – Stratification of a population with respect to two or more variables. Thus, a public opinion poll might cross-stratify a population of voters with respect to income, education, age, and profession. *See also* Stratified Random Sampling

Cumulant – In mathematical statistics, a description of a distribution which is related to its moments. E.11.

Cumulative Distribution – (1) For grouped data, a distribution showing how many of the items are "less than" or "more than" given values (usually the class limits or the class boundaries). The following is a "less than" cumulative distribution corresponding to the frequency distribution on page 44:

Cumulative Distribution—(continued)

Weight (ounces)	*Cumulative Frequency*
less than 15.60	0
less than 15.80	1
less than 16.00	6
less than 16.20	16
less than 16.40	30
less than 16.60	40
less than 16.80	45
less than 17.00	47
less than 17.20	49
less than 17.40	50

(2) For random variables, the term "cumulative distribution" is synonymous with "distribution function." E.3. *See also* Distribution Function

Curve Fitting—(1) The process of describing (approximating) an observed frequency distribution by means of a probability function or a probability density; for example, approximating an observed distribution with a normal curve. This may involve the choice of a particular kind of probability function or probability density, the estimation of its parameters, and the calculation of probabilities corresponding to the classes of the observed distribution. (2) The process of fitting a curve to points representing paired data (or, more generally, a surface to n-tuples of observations). This may involve the choice of a particular kind of curve or surface as well as the estimation of the parameters (or constants) appearing in its equation. *See also* Least Squares, Method of

Curvilinear Correlation—A nonlinear relationship between two or more variables; the strength of such a relationship is sometimes measured by the correlation ratio. *See also* Correlation Ratio

Curvilinear Regression—*See* Regression

Curvilinear Trend—A trend which is nonlinear. *See also* Exponential Trend; Gompertz Curve; Logistic Curve; Polynomial Trend

Cybernetics—A word first suggested (but never precisely defined) by Norbert Wiener. Clearly implied, however, is that all living and mechanical systems are both information systems and control (or feedback) systems. Thus, cybernetics is important from a statistical standpoint since it includes in its scope the statistical uncertainty both of information theory and of the inputs of control systems.

Cycle—In time series analysis, a periodic movement; the *period* or *length* of a cycle is the time it takes for one complete up-and-down and down-and-up movement. *See also* Business Cycle

Cyclical Irregulars – In time series analysis, estimates of the cyclical and irregular components; they are usually obtained by eliminating from a series components attributed to trend and seasonal patterns.

Cyclical Relatives – In time series analysis, quantities (in percentage form) arrived at by removing the trend, and seasonal variation, as well as irregular variations from a series of data.

Cyclical Variation – In time series analysis, that component in a series which presumably results from the action of forces connected with business cycles. *See also* Business Cycle

D

Data – The results of an experiment, census, survey, and any kind of process or operation. *See also* Deseasonalized Data; External Data; Internal Data; Primary Data; Raw Data; Secondary Data

Data Reduction – The process of summarizing large masses of data by the methods of descriptive statistics, namely, by grouping them into tables or representing them by means of statistics such as the mean, a quartile, or the standard deviation.

Death Process, Pure – *See* Pure Death Process

Debugging – The process of locating and correcting errors in a computer routine, or of isolating and eliminating malfunctions of a computer itself.

Deciles – The deciles D_1, D_2, . . . , and D_9 are values at or below which lie, respectively, the lowest 10, 20, . . . , and 90 per cent of a set of data. A.7 and B.5.

Decimal Coding Systems – Schemes, almost always utilizing binary symbols (bits), for representing the decimal digits in a digital computer. In one such widely used system, called *direct binary coding, binary coded decimal,* or the *8421 system,* each decimal digit of a number is separately represented by its binary equivalent; thus, the number 739 is coded 0111 0011 1001. When a binary 3 is added to each digit coded in direct binary, another system, called the *excess-3 system,* results; hence, in excess-3 the number 739 is coded 1010 0110 1100.

Decision Function – *See* Decision Theory

Decision Theory – A unified approach to all problems of estimation, prediction, and hypothesis testing. It is based on the concept of a decision function δ (*delta*), which tells the experimenter how to conduct the statistical aspects of an experiment (for instance, how to choose the size of a sample) and what action to take for each possible outcome. In order to choose a decision function which is

Decision Theory—(continued)
best according to some criterion, one must consider a *loss function* which assigns a numerical value $L(a, \theta)$ (reflecting the cost of experimentation and rewards and penalties for making good, poor, correct, or incorrect decisions) to each pair (a, θ) of actions, a, taken by the experimenter, and values of the parameter θ under consideration. For a given loss function, one can determine a *risk function,* whose values $\rho(\theta, \delta)$ give the *expected loss* to which one is exposed when using a given decision function δ and when the parameter actually has the value θ. *Minimizing the maximum risk* to which one is exposed by the decision functions, or imposing alternate criteria relating to the risk function, one ultimately determines which decision function has the most desirable properties. If the alternate criterion is to *minimize the expected risk,* averaged with respect to θ relative to a given a priori distribution of this parameter, the resulting solution (that is, choice of δ) is referred to as a *Bayes solution;* the corresponding minimum value of the expected risk is referred to as a *Bayes risk.*

Defect—In quality control, a term used to refer to a single instance in which a unit of product fails to conform to some requirement, specification, or standard established for a single quality characteristic. For instance, a ceramic disk which is cracked, off-color, and too thick has *three* defects.

Defective—In quality control, a unit (piece, part, specimen, etc.) which fails to meet a certain requirement. In many applications, a unit is considered to be defective if it has *at least one* defect.

Defining Contrast—In factorial experimentation, a confounded effect which is used to divide a complete factorial experiment into fractional replicates. How this is done is shown in the book by I. Miller and J. E. Freund listed on page 194.

Deflating—In time series analysis, a general term applied to the process of adjusting a value series (reflecting changes in both quantity and price) for changes in price. This is usually done by dividing each number in the value series by an appropriate price index number. For instance, if an index of retail sales (a value series) and an index of retail prices, both with 1959 as base, stood at 124 and 110, respectively, in 1965, then the index of deflated sales, $124/1.10 = 112.7$, is an estimate of the ratio of the quantity of goods sold at retail in 1965 to the quantity sold in 1959.

Degree of Confidence—*See* Confidence Interval

Degree of Truncation—*See* Truncated Data

Degrees of Freedom – (1) A random sample of size n is said to have $n - 1$ degrees of freedom for estimating the population variance, in the sense that there are $n - 1$ independent deviations from the mean on which to base such an estimate. (2) The parameter ν (*nu*) of the chi-square distribution is referred to as its degrees of freedom. (3) The parameter ν of the t distribution is referred to as its degrees of freedom. (4) The parameters ν_1 and ν_2 of the F distribution are referred to as the numerator and denominator degrees of freedom. (5) In applications of a chi-square statistic, the number of degrees of freedom is given by the number of terms in the formula for the statistic *minus* the number of independent restrictions imposed on the expected frequencies; for instance, in the analysis of an r by k contingency table there are $rk - (r + k - 1) = (r - 1)(k - 1)$ degrees of freedom. (6) In an analysis of variance table, the degrees of freedom listed for the various sources of variation are the degrees of freedom of the chi-square distributions of the corresponding mean squares (divided by suitable constants).

Density – *See* Probability Density Function

Density Function – *See* Probability Density Function

Dependent Events – *See* Independent Events

Dependent Random Variables – *See* Independent Random Variables

Dependent Variable – If the value of a function f is given by $y = f(x_1, x_2, \ldots, x_k)$, it is customary to refer to $x_1, x_2, \ldots$, and x_k as the *independent variables* and to y as the *dependent variable*. The major objective of many statistical investigations is to predict values (or expected values) of dependent variables in terms of known or assumed values of independent variables.

Descriptive Statistics – Although this term has been used to refer only to tabular and graphical presentations of statistical data, nowadays it is used more broadly to refer to any treatment of data which does not involve generalizations.

Deseasonalized Data – In time series analysis, monthly (weekly, daily, or hourly) data from which the seasonal variations have been eliminated, as far as possible, by dividing the actual data by corresponding values of a seasonal index (expressed as proportions). For instance, if a store's sales for July, 1965, were $1.4 million and the July seasonal index is 80.0, then the deseasonalized (or *seasonally adjusted*) July, 1965, sales were 1.4/0.80 = $1.75 million. *See also* Annual Rate; Seasonal Index

Design of Experiments – *See* Experimental Design

Destructive Sampling (or Testing) – A testing procedure in which the items sampled are destroyed, damaged, or otherwise altered so

Destructive Sampling (or Testing) – (continued)
that they can no longer be used for the purpose intended. For example, tests of such mechanical properties as tension or shear are destructive.

Determinant – A square array of numbers, called the *elements* of the determinant, symbolizing the sum of certain products of these elements, which arises in the solution of simultaneous linear equations. The number of rows (or columns) is called the *order* of the determinant. For instance, a determinant of order 2 is a square array of the form

$$\begin{vmatrix} a & b \\ c & d \end{vmatrix}$$

and its value is $ad - bc$. Using determinants, the solution of the system of equations

$$ax + by = e$$

$$cx + dy = f$$

can be written

$$x = \frac{\begin{vmatrix} e & b \\ f & d \end{vmatrix}}{\begin{vmatrix} a & b \\ c & d \end{vmatrix}} \qquad y = \frac{\begin{vmatrix} a & e \\ c & f \end{vmatrix}}{\begin{vmatrix} a & b \\ c & d \end{vmatrix}}$$

provided $ad - bc \neq 0$. A determinant of order 3 is a square array of the form

$$\begin{vmatrix} a & b & c \\ d & e & f \\ g & h & i \end{vmatrix}$$

and its value is $aei + bfg + cdh - ceg - bdi - afh$. Further information about determinants may be found in almost any textbook on algebra.

Determination, Coefficient of – The square of the correlation coefficient; it gives the proportion of the total variation of the dependent variable which is accounted for by the linear relationship with the independent variable. **I.20.** The term has also been applied to the square of the multiple correlation coefficient.

Deviation from the Mean – For a given set of data, the amount by which an individual observation differs from the mean; thus, the deviation of the ith observation x_i from the mean $\bar{x}$ is given by $x_i - \bar{x}$. An important property of deviations from the mean is that, for any set of data, their sum is always equal to zero. The absolute value of $x_i - \bar{x}$, namely, $|x_i - \bar{x}|$, is referred to as an *absolute deviation* from the mean.

Dichotomy – A classification which divides the elements of a population (or sample) into two categories. For example, a dichotomy may be a classification into which one groups items as defective or non-defective, or a classification into which one groups individuals as being married or single. A two-way classification where each variable divides the elements of a population (or sample) into two categories is called a *double dichotomy.* For example, a double dichotomy may be a classification in which one groups persons into single males, married males, single females, and married females.

Difference Between Means, Standard Error of – The standard deviation of the sampling distribution of the difference between two sample means. F.4.

Difference Between Proportions, Standard Error of – The standard deviation of the sampling distribution of the difference between two sample proportions. G.3.

Difficulty Value (of an Item) – In psychological and educational test construction, the per cent of a group (often students of a particular age or grade level) who correctly answer the item.

Digital Computer – A computer which, unlike an analog computer, performs the ordinary operations of arithmetic on numbers as such. The main difference between a digital desk calculator and an electronic digital computer is that the latter has an extensive memory for storing information, that it performs long sequences of arithmetical and logical operations without human intervention, and that it operates at a very high speed. Since most formulas, even complex scientific ones, can be reduced to sequences of basic arithmetical operations (at least to any desired degree of approximation), digital computers have a very wide range of applicability. The five *major* components of a computer are: (1) The *input section* which takes information stored externally (as in punched cards or magnetic tape) and "reads" (transfers) it into the internal storage (or memory) of the computer. (2) The *storage* (or *memory*) *section* into which information can be introduced, in which it can be retained, and from which it can be extracted when needed. At present, the most widely used type of internal storage is magnetic core storage, a unit built of many small magnetic cores. Some progress is now being made in

Digital Computer—(continued)

developing thin film and cryogenic (ultracold) memories. (3) The *arithmetic-logic section* in which the arithmetical and logical operations necessary to the solution of a problem are actually performed. (4) The *control section* which causes the various units of the computer to function in such a way that each instruction in storage is sensed at the proper time and executed in the proper manner. (5) The *output section* which transfers information from the internal storage (memory) of the computer to some suitable or desirable external form (punches it in cards, for example, or records it on magnetic tape). The input and output sections are often collectively called the *I/O units*.

Digitizing—The process of converting the analog representation of a quantity (for instance, shaft rotation or voltage) into digital form. Such conversions, made by *analog-to-digital converters*, often involve sampling from the analog quantity at discrete time intervals so as to permit conversion between these instants. In the reverse process, digital representation is converted to analog form by *digital-to-analog converters*.

Discrete Density Function—*See* Probability Function

Discrete Distribution—The distribution of a discrete random variable. *See also* Probability Function

Discrete Random Variable—A random variable whose range is finite or countably infinite. For example, random variables having the binomial or the Poisson distribution are referred to as discrete.

Discriminant Function—A function (associating a number with each possible set of values of one or more selected variables) which is used in conjunction with a set of threshold values in a classification procedure. For example, some linear combination of bone measurements may be used to discriminate between (or separate) two populations of skeletons. Further information may be obtained from the book by P. G. Hoel listed on page 194.

Discriminating Power (of an Item)—In psychological and educational test construction, the extent to which a test item separates those individuals who possess much of a certain trait from those who possess little of the trait.

Dispersion—The extent to which the elements of a sample or the elements of a population are not all alike in size, are spread out, or vary from one another. Measures of this characteristic are usually called *measures of variation*. *See also* Mean Deviation; Standard Deviation

Dispersion, Index of—*See* Binomial Index of Dispersion; Poisson Index of Dispersion

Distribution – (1) For observed data, the term is used to refer to their over-all scattering and also as a synonym for "frequency distribution." (2) The distribution of a random variable is its *probability structure* as described, for example, by its probability function or its probability density function. It is in this sense that one speaks, for example, of a binomial distribution or a normal distribution. *See also* Distribution Function; Frequency Distribution; Probability Density Function; Probability Distribution; Probability Function

Distribution-Free Methods – This term is applied to methods of inference in which no assumptions whatsoever are made about the nature, shape, or form of the populations from which the data are obtained. *See also* Nonparametric Tests

Distribution Function – A function whose values $F(t)$ are the probabilities that a random variable assumes a value less than or equal to t for $-\infty < t < \infty$. E.3. *See also* Joint Distribution Function

Dodge Romig Tables – Tables used extensively in sampling inspection by attributes. They include LTPD (Lot Tolerance Per Cent Defective) Tables for single and double sampling, and AOQL (Average Outgoing Quality Limit) Tables for single and double sampling. Detailed descriptions accompany the tables which are listed on page 193.

Domains of Study – The various subpopulations (strata) of a population whose characteristics (as separate from the characteristics of the total population) are of particular interest. This term is due to the UN Subcommittee on Sampling.

Dominating Strategy – In game theory, one strategy dominates another if there are no circumstances (strategies of the opposing player) for which the second strategy is preferable to the first, and if there is at least one circumstance for which the first strategy is preferable to the second. This concept is analogous to that of one decision function being "uniformly better" than another.

Doolittle Method – A convenient method of solving systems of simultaneous linear equations. The method is illustrated in the book by R. L. Wine listed on page 195. Since its introduction in 1878, there have been various modifications made in the basic procedure. *See also* Wherry-Doolittle (Test Selection) Method

Double Dichotomy – *See* Dichotomy

Double Sampling – *See* Multiple Sampling

Duncan's Multiple Range Test – *See* Multiple Comparisons Tests

Dynamic Programming – A method of finding solutions to multistage decision problems; that is, to problems for whose solution it is required that a sequence of decisions be made. As one of the techniques of mathematical programming, dynamic programming is considerably more flexible than linear programming and, hence,

Dynamic Programming—(continued)
applicable to a more general class of problems. For further information see the book by R. L. Ackoff listed on page 193.

E

Econometrics—That area of economic science in which modern mathematical and statistical techniques are used to analyze economic situations for the purpose of predicting future behavior, or to test the validity of economic theory formulated in mathematical terms.

Edgeworth Index—A weighted aggregative price index in which the weights are the means of the respective quantities produced (consumed, sold, etc.) in the base year and in the given year.

Effect—(1) In factorial experimentation, a quantity (usually a parameter of the model) which represents a change in response produced by a change in level of one or more of the factors. (2) In analysis of variance, a parameter of the so-called fixed-effects model. (**J.1a**) and (**J.3a**). *See also* Block Effect; Interaction; Main Effect; Treatment Effect

Efficiency—*See* Power Efficiency; Relative Efficiency

Efficient—An unbiased estimator is said to be efficient if no other unbiased estimator of the same parameter has a sampling distribution with a smaller variance. An estimator which has this property in the limit, when the sample size becomes infinite, is said to be *asymptotically efficient*. *See also* Relative Efficiency

Elementary Operations—Operations carried out on equations of a given system which lead to another system of equations having the same solutions. Specifically, the elementary operations are (a) interchanging any two equations of a system, (b) multiplying both sides of an equation by a nonzero constant, and (c) adding to both sides of an equation a multiple of the corresponding sides of any other equation of the system. For instance, (1) and (2) below have the same solution, (2) having been derived from (1) by exchanging the two equations, multiplying both sides of the new first equation by 2, and finally adding the respective sides of this equation to those of the other equation.

$$(1)\quad \begin{aligned} 3x + 5y &= 7 \\ 2x + 3y &= 4 \end{aligned} \qquad \text{and (2)} \quad \begin{aligned} 4x + 6y &= 8 \\ 7x + 11y &= 15 \end{aligned}$$

Elementary operations are very useful in solving systems of simultaneous linear equations. *See also* Determinant

Enumeration Data – *See* Count Data

Equal Allocation – In stratified sampling, the allocation of equal parts of the total sample to the individual strata. For instance, if a stratified sample of 100 students is to be taken from among the 400 freshmen, 300 sophomores, 200 juniors, and 100 seniors attending an undergraduate school, equal allocation requires that 25 students be taken from each of the classes (regardless of their differences in size). *See also* Optimum Allocation; Proportional Allocation

Equal Ignorance, Principle of – When using Bayes' formula, the assumption of *equal* a priori probabilities in the absence of any information; often questionable, the principle is always controversial.

Equitable Game – Also called "fair," a game in which each participant has the same expectation; that is, a game which does not favor any player or group of players.

Ergodic Process – A random process is said to be ergodic if averages based on one observed record of the process can serve as estimates of corresponding averages for repeated realizations of the process. A more formal definition may be found in the book by E. Parzen listed on page 195.

Erlang Distribution – An alternate name for the gamma distribution, used mainly in queuing theory. E.35.

Error Mean Square – In analysis of variance, the error sum of squares divided by its degrees of freedom; it provides an estimate of the (supposedly) common error variance of the populations. In the analysis of variance tables (**J.1f**) and (**J.3g**) it is denoted MSE.

Error Rate – In hypothesis testing, the *unconditional* probability of making an error; that is, erroneously accepting or rejecting a hypothesis. Note that the probabilities of Type I and Type II errors are *conditional* probabilities; the first is subject to the condition that the null hypothesis is true, and the second is subject to the condition that the null hypothesis is false.

Error Sum of Squares – In analysis of variance, that component of the total sum of squares which is attributed to experimental error. (**J.1e**) and (**J.3f**).

Error Variance – The variance of a random (or chance) component of a model; the term is used mainly in the presence of other sources of variation, as for example in regression analysis or in analysis of variance. It is referred to as σ^2 in the descriptions of model equations (**J.1a**) and (**J.3a**).

Essential Information – In quality control, that part of the total information contained in an original set of data which, when summarized by such measures as the mean, the standard deviation, or the range,

Essential Information—(continued)

answers whatever questions are asked about the data so fully that further statistical descriptions do not contribute materially to one's knowledge. *See also* Total Information

Estimate—A number or an interval, based on a sample, which is intended to match a parameter of a mathematical model. *See also* Interval Estimation; Point Estimation

Estimator—*See* Point Estimation

Event—In probability theory, an event is a subset of a sample space. Thus, "event" is the nontechnical term and "subset of a sample space" is the corresponding mathematical counterpart. For example, the event of rolling a ten with a pair of dice is the subset which consists of the outcomes where the first die comes up four and the other six, where both dice come up five, and where the first die comes up six and the other four.

Evolutionary Process—*See* Stationary Process

Exhaustive Sampling—This term refers to the 100 per cent inspection of a population.

Expectation—*See* Mathematical Expectation

Expected Frequency—(1) In the analysis of count (or enumeration) data, a cell frequency calculated on the basis of appropriate theory or assumptions; for example, the frequencies calculated according to formula **(G.13b)** for a contingency table. (2) In curve fitting, a class frequency obtained by approximating an observed frequency distribution with a probability function or a probability density function.

Expected Mean Square—In analysis of variance, the expected value of a mean square under a given set of assumptions; that is, for a given model. Expected mean squares serve an important function in setting up significance tests and confidence intervals for the various parameters of the model. See, for example, the book by R. L. Wine listed on page 195.

Expected Utility—*See* Utility Function

Expected Value—The expected value of a random variable is the mean of its distribution. **(E.4a)** and **(E.4c)**. The more general concept of the expected value of a function of a random variable is defined in **(E.4b)** and **(E.4d)**. *See also* Mathematical Expectation

Experimental Condition—In factorial experimentation, the levels at which the individual factors are applied to a given experimental unit. In analysis of variance, the term is used synonymously with "treatment."

Experimental Design—The statistical aspects of the design (or planning) of an experiment are: (a) Selecting the treatments

Experimental Design—(continued)
(factors, and levels of factors) whose effects are to be studied; (b) Specifying a layout for the experimental units (plots) to which the treatments are to be applied; (c) Providing rules according to which the treatments are to be distributed among the experimental units; (d) Specifying what measurements are to be made for each experimental unit. All these things must be accomplished in such a way that the techniques to be used in the analysis of the results are clear prior to the conduct of the experiment.

Experimental Error—The errors, or variations, not accounted for by hypothesis; in analysis of variance, their magnitude is estimated by the error sum of squares. Presumed to be caused by extraneous variables, such errors are often combined under the general heading of "chance variation." Note that in this sense the word "error" does not mean "mistake." *See also* Sampling Error

Experimental Sampling Distribution—A distribution of the values of a statistic obtained by simulating repeated samples from a given population by means of random numbers or other Monte Carlo techniques. For example, if the means of repeated samples from a given population are grouped, the resulting distribution is referred to as an experimental sampling distribution of the mean.

Experimental Unit—In experimental design, an experimental unit is the subject, object, area, grouping, or subdivision to which a treatment is applied. Thus, an experimental unit might be a plot of land, a student taking a given course, several pigs in a pen, a piece of wood, or a batch of seed. Since much of the theory of experimental design was originally developed for agricultural experiments, experimental units are often referred to simply as *plots*. Experimental units have also been described as the smallest divisions of experimental material which are such that any two experimental units may receive different treatments.

Exponential Distribution—Also referred to as the *negative exponential distribution*, this distribution has important applications in engineering to reliability studies and life testing, in queuing theory, and in other areas. E.33. The exponential distribution also arises as the distribution of the waiting times between successive occurrences (arrivals) in a Poisson process.

Exponential Smoothing—A method of forecasting which makes use of exponentially weighted moving averages; it continuously corrects for the amount by which the actual and estimated figures for a just-completed period fail to agree. For further information see the book by R. G. Brown listed on page 193.

Exponential Trend – A trend in a time series which is adequately described by an equation of the form $y = ab^x$. I.2. On semi-logarithmic graph paper such trends appear as straight lines.

Extensive Form of a Game – *See* Finite Game

External Data – Statistical data gathered by an organization from sources outside the organization itself (as by a business firm from government or trade publications).

Extrapolation – The process of estimating (or predicting) a value which lies beyond the range of values on the basis of which the predicting equation was obtained; for example, in estimating 1970 sales from a trend equation fit to sales for the years 1950-1964.

Extreme Values – The smallest and largest values of a sample. These values form the basis of a number of estimation and hypothesis testing procedures. Sometimes, extreme values are analyzed to decide whether values which differ considerably from the bulk of the data can be discarded from the data. *See also* Modified Mean; Outliers

F

***F* Distribution** – A distribution which is of fundamental importance in analysis of variance. E.34. It arises as the sampling distribution of the ratio of the values of two independent random variables having chi-square distributions, each divided by its degrees of freedom; for example, the ratio of the variances of two random samples from normal populations having the same variance. Correspondingly, the parameters ν_1 and ν_2 of the F distribution are referred to as its *numerator* and *denominator degrees of freedom*. The F distribution is also known as the *variance-ratio* distribution. The tables on pages 40 and 41 contain values of F_{α,ν_1,ν_2} for which the area *to their right* under the F distribution with ν_1 and ν_2 degrees of freedom is equal to α.

***F* Test** – A test based on a statistic which (under an appropriate null hypothesis) has an F distribution; for example, F.24, which is a test concerning the equality of two population variances, and the tests described in J.1 and J.3.

Factor – (1) In experimentation, a factor is a variable, or a quantity (a possible source of variation) under investigation. (2) In factor analysis, a factor is a linear combination of (observable) variables to which one attributes a special relevance.

Factor Analysis – In multivariate analysis (especially in psychological applications), a method of expressing data linearly in terms of factors which are of special relevance so far as the construction of appropriate models is concerned. For example, the scores n individuals obtained on k tests may be related linearly to such relevant factors as arith-

Factor Analysis—(continued)
metic or verbal facility. For a detailed discussion of this subject see the book by R. B. Cattell listed on page 193.

Factor-Reversal Test—One of several mathematical tests of index number quality, the factor-reversal test requires that the product of a price index and a corresponding quantity index (obtained by replacing prices with quantities and vice versa) equal a corresponding value index.

Factorial Experiment—A *complete* factorial experiment is an experiment in which all levels of each factor (variable) are investigated in combination with all levels of every other factor. It is customary to denote complete factorial experiments by indicating the number of levels for each factor; thus, a 3×4 factorial experiment is a two-factor experiment where one factor has three levels and the other has four levels. Similarly, a 2^3 factorial is a three-factor experiment where each factor has two levels. If certain properly chosen levels of factors are omitted, the experiment is referred to as a *fractional factorial experiment,* or as a *fractional replicate* of a factorial experiment. Usually, the experimental conditions included in a fractional replicate of a factorial experiment are chosen in such a way that main effects and some of the lower-order interactions can be estimated, while higher-order interactions are confounded with each other. Many of the methods of *response surface analysis* (for example, methods utilizing rotatable designs) are also included under the general heading of factorial experiments. *See also* Confounding; Defining Contrast

Factorial Notation—In mathematics, the product of all positive integers less than or equal to the positive integer n is referred to as "n factorial" and it is denoted $n!$. Thus, $3! = 3 \cdot 2 \cdot 1 = 6$, $4! = 4 \cdot 3 \cdot 2 \cdot 1 = 24$, $5! = 5 \cdot 4 \cdot 3 \cdot 2 \cdot 1 = 120$, and so on. Also, by definition $0! = 1$. A more general definition is given in terms of the *gamma function* in **(E.35b)**.

Failure Rate—In reliability studies in engineering, the instantaneous rate at which items fail after they have survived to a given time t.

Failure-Time Distribution—In reliability studies in engineering, it is given by the probability density function of an item failing (for the first time) at time t; in the discrete case it is given by the probability function of an item failing (for the first time) on the kth trial.

Fiducial Inference—*See* Fiducial Probability

Fiducial Limits—Limits similar to confidence limits, based on the concept of fiducial probability. Whereas in the theory of confidence intervals the endpoints of the interval are the random variables, in the construction of fiducial limits the parameter itself is assumed to have a (fiducial) distribution. Some examples of fiducial limits are given in the book by R. A. Fisher listed on page 194.

THE F DISTRIBUTION (Values of $F_{.05,\nu_1,\nu_2}$)*

The entries in this table are values of $F_{.05,\nu_1,\nu_2}$, for which the area to their right under the F distribution with ν_1 and ν_2 degrees of freedom is equal to 0.05.

ν_1 = Degrees of freedom for numerator

ν_2 = Degrees of freedom for denominator	1	2	3	4	5	6	7	8	9	10	12	15	20	24	30	40	60	120	∞
1	161	200	216	225	230	234	237	239	241	242	244	246	248	249	250	251	252	253	254
2	18.5	19.0	19.2	19.2	19.3	19.3	19.4	19.4	19.4	19.4	19.4	19.4	19.4	19.5	19.5	19.5	19.5	19.5	19.5
3	10.1	9.55	9.28	9.12	9.01	8.94	8.89	8.85	8.81	8.79	8.74	8.70	8.66	8.64	8.62	8.59	8.57	8.55	8.53
4	7.71	6.94	6.59	6.39	6.26	6.16	6.09	6.04	6.00	5.96	5.91	5.86	5.80	5.77	5.75	5.72	5.69	5.66	5.63
5	6.61	5.79	5.41	5.19	5.05	4.95	4.88	4.82	4.77	4.74	4.68	4.62	4.56	4.53	4.50	4.46	4.43	4.40	4.37
6	5.99	5.14	4.76	4.53	4.39	4.28	4.21	4.15	4.10	4.06	4.00	3.94	3.87	3.84	3.81	3.77	3.74	3.70	3.67
7	5.59	4.74	4.35	4.12	3.97	3.87	3.79	3.73	3.68	3.64	3.57	3.51	3.44	3.41	3.38	3.34	3.30	3.27	3.23
8	5.32	4.46	4.07	3.84	3.69	3.58	3.50	3.44	3.39	3.35	3.28	3.22	3.15	3.12	3.08	3.04	3.01	2.97	2.93
9	5.12	4.26	3.86	3.63	3.48	3.37	3.29	3.23	3.18	3.14	3.07	3.01	2.94	2.90	2.86	2.83	2.79	2.75	2.71
10	4.96	4.10	3.71	3.48	3.33	3.22	3.14	3.07	3.02	2.98	2.91	2.85	2.77	2.74	2.70	2.66	2.62	2.58	2.54
11	4.84	3.98	3.59	3.36	3.20	3.09	3.01	2.95	2.90	2.85	2.79	2.72	2.65	2.61	2.57	2.53	2.49	2.45	2.40
12	4.75	3.89	3.49	3.26	3.11	3.00	2.91	2.85	2.80	2.75	2.69	2.62	2.54	2.51	2.47	2.43	2.38	2.34	2.30
13	4.67	3.81	3.41	3.18	3.03	2.92	2.83	2.77	2.71	2.67	2.60	2.53	2.46	2.42	2.38	2.34	2.30	2.25	2.21
14	4.60	3.74	3.34	3.11	2.96	2.85	2.76	2.70	2.65	2.60	2.53	2.46	2.39	2.35	2.31	2.27	2.22	2.18	2.13
15	4.54	3.68	3.29	3.06	2.90	2.79	2.71	2.64	2.59	2.54	2.48	2.40	2.33	2.29	2.25	2.20	2.16	2.11	2.07
16	4.49	3.63	3.24	3.01	2.85	2.74	2.66	2.59	2.54	2.49	2.42	2.35	2.28	2.24	2.19	2.15	2.11	2.06	2.01
17	4.45	3.59	3.20	2.96	2.81	2.70	2.61	2.55	2.49	2.45	2.38	2.31	2.23	2.19	2.15	2.10	2.06	2.01	1.96
18	4.41	3.55	3.16	2.93	2.77	2.66	2.58	2.51	2.46	2.41	2.34	2.27	2.19	2.15	2.11	2.06	2.02	1.97	1.92
19	4.38	3.52	3.13	2.90	2.74	2.63	2.54	2.48	2.42	2.38	2.31	2.23	2.16	2.11	2.07	2.03	1.98	1.93	1.88
20	4.35	3.49	3.10	2.87	2.71	2.60	2.51	2.45	2.39	2.35	2.28	2.20	2.12	2.08	2.04	1.99	1.95	1.90	1.84
21	4.32	3.47	3.07	2.84	2.68	2.57	2.49	2.42	2.37	2.32	2.25	2.18	2.10	2.05	2.01	1.96	1.92	1.87	1.81
22	4.30	3.44	3.05	2.82	2.66	2.55	2.46	2.40	2.34	2.30	2.23	2.15	2.07	2.03	1.98	1.94	1.89	1.84	1.78
23	4.28	3.42	3.03	2.80	2.64	2.53	2.44	2.37	2.32	2.27	2.20	2.13	2.05	2.01	1.96	1.91	1.86	1.81	1.76
24	4.26	3.40	3.01	2.78	2.62	2.51	2.42	2.36	2.30	2.25	2.18	2.11	2.03	1.98	1.94	1.89	1.84	1.79	1.73
25	4.24	3.39	2.99	2.76	2.60	2.49	2.40	2.34	2.28	2.24	2.16	2.09	2.01	1.96	1.92	1.87	1.82	1.77	1.71
30	4.17	3.32	2.92	2.69	2.53	2.42	2.33	2.27	2.21	2.16	2.09	2.01	1.93	1.89	1.84	1.79	1.74	1.68	1.62
40	4.08	3.23	2.84	2.61	2.45	2.34	2.25	2.18	2.12	2.08	2.00	1.92	1.84	1.79	1.74	1.69	1.64	1.58	1.51
60	4.00	3.15	2.76	2.53	2.37	2.25	2.17	2.10	2.04	1.99	1.92	1.84	1.75	1.70	1.65	1.59	1.53	1.47	1.39
120	3.92	3.07	2.68	2.45	2.29	2.18	2.09	2.02	1.96	1.91	1.83	1.75	1.66	1.61	1.55	1.50	1.43	1.35	1.25
∞	3.84	3.00	2.60	2.37	2.21	2.10	2.01	1.94	1.88	1.83	1.75	1.67	1.57	1.52	1.46	1.39	1.32	1.22	1.00

*Reproduced from M. Merrington and C. M. Thompson, "Tables of percentage points of the inverted beta (F) distribution," *Biometrika*, Vol. 33 (1943), by permission of the *Biometrika* trustees.

THE F DISTRIBUTION (Values of $F_{.01,\nu_1,\nu_2}$)*

The entries in this table are values of $F_{.01,\nu_1,\nu_2}$, for which the area to their right under the F distribution with ν_1 and ν_2 degrees of freedom is equal to 0.01.

ν_1 = Degrees of freedom for numerator

ν_2 = Degrees of freedom for denominator	1	2	3	4	5	6	7	8	9	10	12	15	20	24	30	40	60	120	∞
1	4,052	5,000	5,403	5,625	5,764	5,859	5,928	5,982	6,023	6,056	6,106	6,157	6,209	6,235	6,261	6,287	6,313	6,339	6,366
2	98.5	99.0	99.2	99.2	99.3	99.3	99.4	99.4	99.4	99.4	99.4	99.4	99.4	99.5	99.5	99.5	99.5	99.5	99.5
3	34.1	30.8	29.5	28.7	28.2	27.9	27.7	27.5	27.3	27.2	27.1	26.9	26.7	26.6	26.5	26.4	26.3	26.2	26.1
4	21.2	18.0	16.7	16.0	15.5	15.2	15.0	14.8	14.7	14.5	14.4	14.2	14.0	13.9	13.8	13.7	13.7	13.6	13.5
5	16.3	13.3	12.1	11.4	11.0	10.7	10.5	10.3	10.2	10.1	9.89	9.72	9.55	9.47	9.38	9.29	9.20	9.11	9.02
6	13.7	10.9	9.78	9.15	8.75	8.47	8.26	8.10	7.98	7.87	7.72	7.56	7.40	7.31	7.23	7.14	7.06	6.97	6.88
7	12.2	9.55	8.45	7.85	7.46	7.19	6.99	6.84	6.72	6.62	6.47	6.31	6.16	6.07	5.99	5.91	5.82	5.74	5.65
8	11.3	8.65	7.59	7.01	6.63	6.37	6.18	6.03	5.91	5.81	5.67	5.52	5.36	5.28	5.20	5.12	5.03	4.95	4.86
9	10.6	8.02	6.99	6.42	6.06	5.80	5.61	5.47	5.35	5.26	5.11	4.96	4.81	4.73	4.65	4.57	4.48	4.40	4.31
10	10.0	7.56	6.55	5.99	5.64	5.39	5.20	5.06	4.94	4.85	4.71	4.56	4.41	4.33	4.25	4.17	4.08	4.00	3.91
11	9.65	7.21	6.22	5.67	5.32	5.07	4.89	4.74	4.63	4.54	4.40	4.25	4.10	4.02	3.94	3.86	3.78	3.69	3.60
12	9.33	6.93	5.95	5.41	5.06	4.82	4.64	4.50	4.39	4.30	4.16	4.01	3.86	3.78	3.70	3.62	3.54	3.45	3.36
13	9.07	6.70	5.74	5.21	4.86	4.62	4.44	4.30	4.19	4.10	3.96	3.82	3.66	3.59	3.51	3.43	3.34	3.25	3.17
14	8.86	6.51	5.56	5.04	4.70	4.46	4.28	4.14	4.03	3.94	3.80	3.66	3.51	3.43	3.35	3.27	3.18	3.09	3.00
15	8.68	6.36	5.42	4.89	4.56	4.32	4.14	4.00	3.89	3.80	3.67	3.52	3.37	3.29	3.21	3.13	3.05	2.96	2.87
16	8.53	6.23	5.29	4.77	4.44	4.20	4.03	3.89	3.78	3.69	3.55	3.41	3.26	3.18	3.10	3.02	2.93	2.84	2.75
17	8.40	6.11	5.19	4.67	4.34	4.10	3.93	3.79	3.68	3.59	3.46	3.31	3.16	3.08	3.00	2.92	2.83	2.75	2.65
18	8.29	6.01	5.09	4.58	4.25	4.01	3.84	3.71	3.60	3.51	3.37	3.23	3.08	3.00	2.92	2.84	2.75	2.66	2.57
19	8.19	5.93	5.01	4.50	4.17	3.94	3.77	3.63	3.52	3.43	3.30	3.15	3.00	2.92	2.84	2.76	2.67	2.58	2.49
20	8.10	5.85	4.94	4.43	4.10	3.87	3.70	3.56	3.46	3.37	3.23	3.09	2.94	2.86	2.78	2.69	2.61	2.52	2.42
21	8.02	5.78	4.87	4.37	4.04	3.81	3.64	3.51	3.40	3.31	3.17	3.03	2.88	2.80	2.72	2.64	2.55	2.46	2.36
22	7.95	5.72	4.82	4.31	3.99	3.76	3.59	3.45	3.35	3.26	3.12	2.98	2.83	2.75	2.67	2.58	2.50	2.40	2.31
23	7.88	5.66	4.76	4.26	3.94	3.71	3.54	3.41	3.30	3.21	3.07	2.93	2.78	2.70	2.62	2.54	2.45	2.35	2.26
24	7.82	5.61	4.72	4.22	3.90	3.67	3.50	3.36	3.26	3.17	3.03	2.89	2.74	2.66	2.58	2.49	2.40	2.31	2.21
25	7.77	5.57	4.68	4.18	3.86	3.63	3.46	3.32	3.22	3.13	2.99	2.85	2.70	2.62	2.53	2.45	2.36	2.27	2.17
30	7.56	5.39	4.51	4.02	3.70	3.47	3.30	3.17	3.07	2.98	2.84	2.70	2.55	2.47	2.39	2.30	2.21	2.11	2.01
40	7.31	5.18	4.31	3.83	3.51	3.29	3.12	2.99	2.89	2.80	2.66	2.52	2.37	2.29	2.20	2.11	2.02	1.92	1.80
60	7.08	4.98	4.13	3.65	3.34	3.12	2.95	2.82	2.72	2.63	2.50	2.35	2.20	2.12	2.03	1.94	1.84	1.73	1.60
120	6.85	4.79	3.95	3.48	3.17	2.96	2.79	2.66	2.56	2.47	2.34	2.19	2.03	1.95	1.86	1.76	1.66	1.53	1.38
∞	6.63	4.61	3.78	3.32	3.02	2.80	2.64	2.51	2.41	2.32	2.18	2.04	1.88	1.79	1.70	1.59	1.47	1.32	1.00

*Reproduced from M. Merrington and C. M. Thompson, "Tables of percentage points of the inverted beta (F) distribution," *Biometrika*, Vol. 33 (1943), by permission of the *Biometrika* trustees.

Fiducial Probability—A concept of probability, due to R. A. Fisher, which is applied to the parameter of populations about which inferences are to be made. Accordingly, such inferences are referred to as *fiducial inferences.* A discussion of this concept of probability may be found in the book by R. A. Fisher listed on page 194.

Finite Game—A game with a finite number of moves, each of which involves a finite number of alternatives. A finite zero-sum two-person game is said to be *rectangular* if each player has only one move. Any game that is expressed as a rectangular game is said to be in *normal form;* otherwise, it is said to be in *extensive form.*

Finite Population—A well-defined set consisting of a finite number of elements. The term is used mainly when a sample of some of these elements forms the basis of an inference concerning the entire set.

Fisher's Ideal Index—An index number which meets various mathematical tests of quality and attempts to eliminate the upward bias of the Laspeyres' index and the downward bias of the Paasche Index by taking the geometric mean of the two. **K.11.**

Fixed-Base Index—An index number series in which all comparisons are made to the same base year (or period).

Fixed-Effects Model—The model used in analysis of variance when the treatment effects, block effects, etc., are looked upon as parameters, namely, as constants. This model is also referred to as Model I. *See also* Mixed Model; Random-Effects Model

Fixed-Weight Aggregative Index—An index number (currently in favor) in which the prices are weighted by corresponding quantities referring to some (fixed) period other than the base year or the given year. **K.8.**

Floating Point Arithmetic—A modification of the system of scientific notation designed to facilitate arithmetic calculations in digital computers. For example, the number 53.42, or $0.5342(10^2)$, could be represented in the IBM 1620 as a six-digit floating point number $\bar{5}342\bar{0}2$. Here the first four digits are the "mantissa," the last two are the "exponent," and the "flags" over the 5 and 0 separate the two parts by marking the high-order digit in each. Use of floating point arithmetic relieves the programmer of the (sometimes almost impossible) task of keeping track of the decimal point in the complex and lengthy machine computations which are often required in scientific data processing.

Flowchart—A chart or diagram on which certain conventions, symbols, and abbreviations are used to outline the elementary steps, and the order in which they are to be performed, in the logical analysis of a problem to be applied to a digital computer. For instance, the flowchart of the accompanying figure shows the logical steps, and their order, in a program to cause a computer to calculate the square of 10

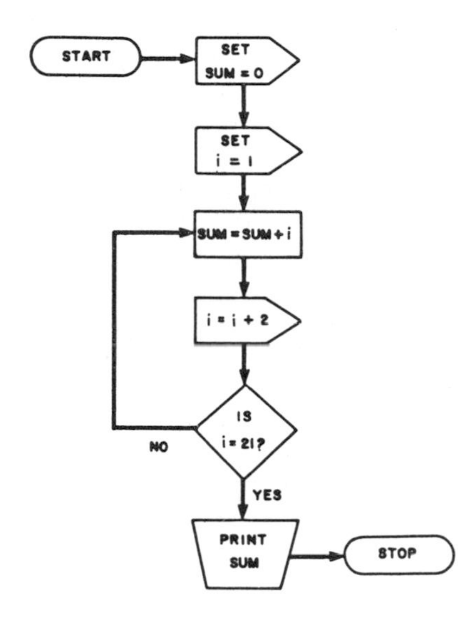

A Flowchart

Flowchart—(continued)
by adding the odd numbers 1, 3, 5, . . . , 19, print it, then halt. Analogous to the flowchart for a digital computer is the "road map" used in connection with an analog computer.

Forecasting—Predictions that involve explaining events which will occur at some future time are called *forecasts*, and the process of arriving at such explanations is called *forecasting*. In so-called *intrinsic* methods of prediction, predictions of the future values of variables are based on their past values; this includes most of the standard statistical methods used in time series analysis.

FORTRAN (FORmula TRANslation System)—A computer programming system, consisting of a mathematically oriented language, largely independent of the computer on which a FORTRAN program is to be executed, together with a compiler for converting a source program written in the FORTRAN language into an object program which can be executed by a particular computer. The mathematical nature of the FORTRAN language is clear from the following two statements, instructing a computer to find the roots of the quadratic equation $Ax^2 + Bx + C = 0$:

```
ROOT1 = (-B+SQRTF(B**2-4.*A*C))/(2.*A)
ROOT2 = (-B-SQRTF(B**2-4.*A*C))/(2.*A)
```

FORTRAN (FORmula TRANslation System)—(continued)

Used initially with the IBM 704, FORTRAN, now available in several versions, is the nearest thing to a "universal" computer programming language in existence.

Fourfold Point Correlation—A synonym for "phi coefficient." G.17.

Fractile—A value, also called a *quantile,* at or below which lies a given fraction (1/10, fifty per cent, 3/4, etc.) of a set of data. *See also* Deciles; Median; Percentiles; Quartiles

Fraction Defective—In quality control, the proportion of defective units in a sample or population. A *fraction-defective chart,* or *p chart,* is a control chart on which one plots the proportion of defectives observed in samples taken at regular intervals of time. L.11 and L.12.

Fractional Factorial—*See* Factorial Experiment

Fractional Replicate—*See* Factorial Experiment

Frequency—The number of items, or cases, falling (or expected to fall) into a category or classification. *See also* Expected Frequency; Observed Frequency

Frequency Distribution—A table (or other kind of arrangement) which shows the classes into which a set of data has been grouped together with the corresponding frequencies, that is, the number of items falling into each class. The following is a frequency distribution of the weights of fifty boxes of detergent taken from the production of a filling machine:

Weight (ounces)	*Number of Boxes*
15.60 - 15.79	1
15.80 - 15.99	5
16.00 - 16.19	10
16.20 - 16.39	14
16.40 - 16.59	10
16.60 - 16.79	5
16.80 - 16.99	2
17.00 - 17.19	2
17.20 - 17.39	1
	50

Other items of interest in connection with this frequency distribution (see the individual listings) are: (1) the *class limits,* which are 15.60 and 15.79, 15.80 and 15.99, . . . , and 17.20 and 17.39; (2) the *class boundaries,* which are 15.595, 15.795, 15.995, . . . , 17.195, and 17.395; (3) the *class marks,* 15.695, 15.895, . . . , and 17.295; and (4) the *class interval,* 0.20. *See also* Cumulative Distribution; Frequency Polygon; Histogram; Ogive; Percentage Distribution

Frequency Function—*See* Probability Density Function; Probability Function

Frequency Interpretation of Probability—A theory in which the probability of an event is interpreted as the proportion of the time the event will occur in the long run. Accordingly, probabilities are estimated by sample proportions; for instance, the probability that a patient will survive a given sickness is estimated by the proportion of recoveries from the sickness that have been observed in the past.

Frequency Polygon—The graph of a frequency distribution obtained by drawing straight lines joining successive points representing the class frequencies, plotted at the corresponding class marks. To complete the picture, classes with zero frequencies are usually added at both ends of a distribution.

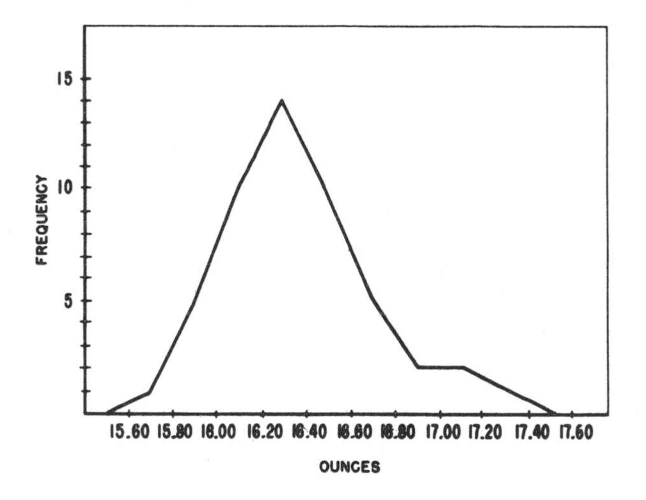

Frequency Polygon of the Weight Distribution on page 44

Frequency Table—A tabular presentation of a frequency distribution; this term is often used interchangeably with "frequency distribution."

Friedman Test—A nonparametric test applicable to the analysis of randomized block designs, namely, to the analysis of k sets of matched observations, where one observation in each set corresponds to a given treatment. The observations are ranked separately within each set and the test statistic is based on the sums of the ranks assigned to the individual treatments. A discussion of this test may be found in the book by S. Siegel listed on page 195.

G

Gambler's Ruin – The problem of two players who play a series of games and bet on the outcome of each game until the capital of one of them is exhausted. Variations of this kind of problem arise in the theory of random walk and in sequential sampling.

Game of Strategy – In contrast to games of chance where the outcome depends only on chance, games of strategy are games in which the outcome depends also, or entirely, on the moves (strategies) chosen by the individual players. For example, whereas *heads or tails* is a game of chance, *poker* is a game of strategy. Note that the word "game" is used here in a very wide sense, applying to any competitive situation involving two or more individuals or parties; for instance, it is used in connection with the conduct of a war, economic planning, or competition in business. A game in which there are *n* players (or groups of players) with conflicting interests is called an *n-person game*. Also, a game in which there is no creation or loss of capital is referred to as a *zero-sum game;* in other words, whatever amount is won (or lost) by some participants (players or groups of players) in a zero-sum game must be lost (or won) by the other participants in the game. Much of the theory that has been developed in connection with games pertains to *zero-sum two-person games. See also* Continuous Game; Finite Game

Game Theory – The mathematical theory which pertains to the determination of optimum strategies in a competitive situation (game of strategy) involving two or more individuals or parties. A general introduction to the theory of games is given in the book by J. C. C. McKinsey listed on page 194. *See also* Minimax Criterion; Strategy

Gamma Distribution – A distribution which includes, as special cases, the chi-square distribution and the exponential distribution. **E.35.** It has many important applications; in Bayesian inference, for example, it is sometimes used as the a priori distribution for the parameter (mean) of a Poisson distribution.

Gauss-Markov Theorem – An important theorem in the theory of estimation; in its simplest form, the theorem asserts that (under certain general conditions) among all unbiased linear estimates of regression coefficients the ones obtained by the method of least squares have a minimum variance. For further information see the book by S. S. Wilks listed on page 195.

Gaussian Distribution – *See* Normal Distribution

General Purpose Index – An index number which attempts to measure changes in such broad and complicated phenomena as production, wholesale prices, or consumer prices. Index numbers which measure

General Purpose Index—(continued)
changes in somewhat more limited phenomena are sometimes called *special purpose indices.*

General Rule of Addition—In probability theory, a formula for calculating the probability that at least one of two events will occur. (D.7a). When the two events are mutually exclusive, the corresponding formula is sometimes referred to as the *Special Rule of Addition,* which is, in fact, postulate (D.1c).

General Rule of Multiplication—In probability theory, a formula for calculating the probability that two events will both occur. (D.8a). When the two events are independent, the corresponding formula is sometimes referred to as the *Special Rule of Multiplication,* which is given by the formula in D.3.

Generalized Interaction—The interaction between two confounded effects. In a 2^n factorial experiment, the generalized interaction between two confounded effects is obtained by "multiplying" their symbolic representations and crossing out repeated letters. For example, the generalized interaction of the confounded *ABCD* and *BDE* interactions is the $A\not{B}C\not{D}\not{B}\not{D}E$, namely, the *ACE* interaction. In any experiment, the generalized interaction of two confounded effects is also confounded.

Geometric Distribution—A probability function whose successive values are in geometric progression. E.25. It arises, for example, as the distribution for the first success in a series of Bernoulli trials.

Geometric Mean—A special kind of "average;" for a set of n positive numbers it is given by the nth root of their product and, unless the numbers are all alike, it is always less than their arithmetic mean. A.3 and B.2. The geometric mean of the numbers 4, 6, and 9, for example, is $(4 \cdot 6 \cdot 9)^{1/3} = 6$. Geometric means are used mainly for averaging rates of change (for instance, index numbers); in practice, their calculation is facilitated by the fact that the logarithm of a geometric mean equals the corresponding arithmetic mean of the logarithms of the numbers. (A.3b) and (B.2b).

Given Year (or Given Period)—In index number construction, the year (or period) one wants to compare. It is usually denoted by the subscript "n;" for example, the price of a commodity in the given year is written p_n and the corresponding quantity produced (consumed, sold, etc.) is written q_n.

Gompertz Curve—One of several growth curves having the general shape of an elongated letter *S*. I.7. It is used to describe the underlying movement of series in which the rate of growth is small at first, increases over a period of time, and then slows as the series levels off.

Goodness of Fit—(1) In the comparison of observed frequencies and expected frequencies, the closeness of the agreement between the two sets of frequencies; it is usually measured by an appropriate chi-square statistic. **G.19.** (2) In fitting a curve to paired data (or a surface to triples of numbers, etc.), the closeness of the points to the curve (surface, etc.); it is measured by various criteria, including the correlation coefficient **I.17** when fitting a straight line to paired data. *See also* Kolmogorov-Smirnov Tests; Least Squares, Method of

Grand Mean—In analysis of variance models, the parameter μ which is estimated by the mean of all the observations in an experiment. **(J.1a)** and **(J.3a)**. The term has also been used to denote the mean of all the observations in an experiment itself, and the over-all mean of several sets of data.

Graphical Presentation—The presentation of statistical data in graphical form, including line charts, bar charts, pie charts, pictograms, maps, and the like.

Greco-Latin Square—(1) In mathematics, a square array of Roman and Greek letters, where each letter (Roman and Greek) appears once in each row and in each column, and each Roman letter appears once in combination with each Greek letter. If one considers the Roman and Greek letters of a Greco-Latin square as constituting separate squares, one refers to these squares as *orthogonal Latin squares.* (2) In statistics, a Greco-Latin square is an experimental design consisting of an array of experimental conditions having the above properties. Such a design permits the simultaneous study of four sources of variation: variables represented, respectively, by the rows; the columns; the Roman letters; and the Greek letters. The following is an example of a 3 by 3 Greco-Latin square:

A β	B γ	C α
C γ	A α	B β
B α	C β	A γ

Greek Alphabet—*See inside of front cover*

Grouped Data—A set of data which has been grouped, or classified, according to some quantitative or qualitative characteristic; that is, data which have been put into a frequency distribution.

Grouping Error—*See* Sheppard's Correction

Growth Curve—*See* Gompertz Curve; Logistic Curve; Modified Exponential Curve

H

H Test – *See* Kruskal-Wallis Test

Hardware (of a Computer) – The various mechanical, electrical, electronic, and magnetic devices (wires, resistors, relays, transistors, connectors, frames, etc.) with which a computer is built. *See also* Software

Harmonic Analysis – The analysis of a series of values into a sum of trigonometric terms of varying period and amplitude. In time series work, harmonic analysis is sometimes used to smooth the irregular movements out of the cyclical irregulars. *See also* Periodogram

Harmonic Mean – A rarely used special "average;" for a set of n numbers it is given by n divided by the sum of their reciprocals. A.4. For instance, if one spends \$12 for parts costing 40 cents a dozen, and another \$12 for parts costing 60 cents a dozen, the average price per dozen is given by the harmonic mean, viz., $\frac{2}{1/40 + 1/60} = 48$ cents.

Heuristic (Computer) Program – A computer program which performs certain complex information processing tasks by using the same sort of selectivity in searching for solutions that human beings use. Thus, a program which emulates certain aspects of human thinking. Programs of this sort have been written to prove theorems in Euclidean geometry, to discover proofs of theorems in logic, etc.

Hierarchial Experiment – *See* Nested Experiment

Higher-Order Interaction – *See* Interaction

Histogram – A graph of a frequency distribution obtained by drawing rectangles whose bases coincide with the class intervals and whose

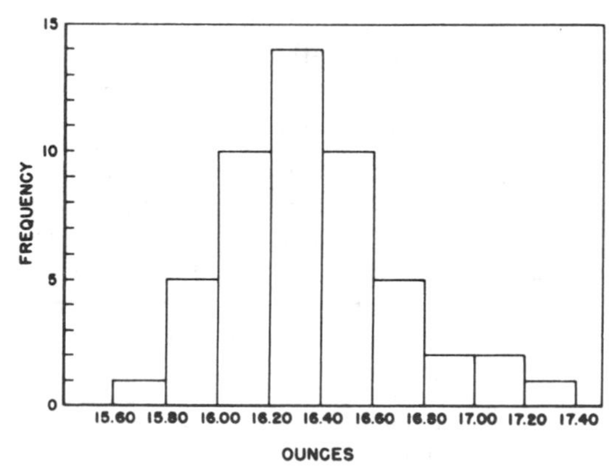

Histogram of the Weight Distribution on page 44

Histogram—(continued)

areas are proportional to the class frequencies. In a histogram representing a distribution with *equal* classes, the ***heights*** of the rectangles are also proportional to the class frequencies.

Homogeneity of Variances—*See* Bartlett's Test

Homoscedasticity—In regression analysis, the property that the conditional distributions of Y for fixed values of the independent variable all have the same variance.

Hybrid Computing System—A computing system which combines analog and digital techniques and equipment in such a way as to realize some of the main advantages of both systems (for instance, the speed of analog machines, and the ability of digital machines to carry out long sequences of complex logical operations). Hybrid systems are essentially of four types: a digital and an analog computer linked together; a digital computer with some analog elements added; an analog computer with some digital elements added; and a system comprised of analog and digital elements (but not including an analog or digital computer intended for independent operation). Hybrid systems are now being used widely in performing engineering simulations of various kinds, and in signal processing.

Hypergeometric Distribution—A distribution which applies to sampling without replacement from a finite population. E.26. If a population consists of a elements of one kind and $N - a$ elements of another kind, the probability function of the hypergeometric distribution gives the probability of getting x elements of the first kind in a random sample of size n. When N is large and n is small compared to N, the hypergeometric distribution is usually approximated by means of the binomial distribution. Hypergeometric distributions are widely used in quality control and sample survey analysis.

Hypothesis—In statistics, an assertion about the parameter (or parameters) of a population or an assertion about the functional form of a population. For example, one might formulate and test the hypothesis that a sample comes from a population with the mean μ_0, or one might assert and test the hypothesis that a sample comes from a normal population. *See also* Alternative Hypothesis; Composite Hypothesis; Null Hypothesis; Simple Hypothesis

Hypothesis Testing—*See* Hypothesis; Neyman-Pearson Theory; Significance Test; Tests of Hypotheses

I

Ideal Index — *See* Fisher's Ideal Index

Imputation — A distribution of the total winnings in an n-person game which is such that: (1) each player receives at least as much as he could make by himself (that is, without participating in a coalition), and (2) the sum of the amounts received by the n players equals the amount the players together could assure for themselves if they all cooperated.

Incomplete Block Design — An experimental design in which all treatments are not represented in each block. *See also* Balanced Design; Partially Balanced Incomplete Block Design

Incomplete Latin Square — *See* Youden Square

Independent Events — In probability theory, two events are said to be independent if, and only if, the probability that they will both occur equals the product of the probabilities that each one, individually, will occur. **D.3.** If two events are not independent, they are said to be *dependent.*

Independent Random Variables — Two or more random variables are independent if, and only if, the values of their joint distribution function are given by the products of the corresponding values of their individual (marginal) distribution functions. If random variables are not independent, they are said to be dependent.

Independent Samples — Two samples are independent if the selection of one in no way affects the selection of the other. More rigorously, two samples of size n_1 and n_2 are independent if they consist of the values of $n_1 + n_2$ independent random variables, with the first n_1, and the other n_2 having, respectively, identical distributions. For example, two random samples of the IQ s of students selected separately at two universities are independent, whereas the IQ s of husbands and their wives do not constitute independent random samples. Samples which are not independent are referred to as *correlated, paired,* or *matched.*

Independent Variable — *See* Dependent Variable

Index Number — A statistical measure which reflects a comparison of prices, quantities, or values pertaining to two different periods of time, two different locations, etc. Many important index numbers which reflect changes in various economic variables are published *in series,* that is, at regular intervals of time, by the Federal government and other organizations.

Index of Seasonal Variation—*See* Seasonal Index

Inductive Statistics—*See* Statistical Inference

Industrial Production, Index of—Constructed and published since the 1920 s by the Federal Reserve Board, the Index of Industrial Production is intended to measure changes in the physical volume, or quantity, of the output of this nation's factories, mines, and electric and gas utilities. The primary source of this important index is the *Federal Reserve Bulletin.*

Inference—*See* Bayesian Inference; Fiducial Probability; Statistical Inference

Infinite Game—A game in which at least one player has infinitely many pure strategies.

Infinite Population—(1) The infinite set of values which can be assumed by a continuous random variable. (2) The phrase "sampling from an infinite population" refers to the process of obtaining a sample which consists of the values of independent random variables having the same (population) distribution. In contrast to sampling *without replacement* from a finite population, the composition of an infinite population is *not* affected by values previously drawn.

Information—A theoretical concept, due to R. A. Fisher, which is closely related to that of the efficiency, or relative efficiency, of an estimator. For further information see the book by R. A. Fisher listed on page 194.

Information Theory—A recently developed mathematical theory of communication involving the application of probability theory; it is concerned with the transmission of messages when the elements of information comprising the message are subject to random disturbances, distortion, transmission failure, and other sources of interference.

Input-Output Accounts—A set of interindustry accounts constructed by the Federal government (using input-output analysis) as a conceptually integrated component of the national income and product accounts. The comprehensive system of integrated national accounts is intended to promote understanding of the interaction between the various industries and final markets of the nation's economy.

Input-Output Analysis—A type of interindustry analysis, that is, a method of determining the various interrelations that arise from production. Input-output analysis is a particular form of linear programming and it constitutes an area of major importance in linear economics. Since it provides a series of connections between the demands of final markets and the output of industry, input-output analysis is a powerful tool for analyzing changes in the economy. The basic model,

Input-Output Analysis—(continued)
of which there are now various extensions and generalizations, was provided by W. Leontief.

Input Section (of a Digital Computer)—*See* Digital Computer

Integer Programming—A technique for solving certain mathematical programming problems which require that in the solution all values of the variables which are to be determined must be integers (or whole numbers).

Interaction—In general, a joint effect of several variables or factors. In factorial experimentation, it is a measure of the extent to which a change in response produced by changes in the levels of one or more factors depends on the levels of the other factors. Interactions involving at least three factors (at least four, as the case may be, etc.) are referred to as *higher-order interactions*, while the others are referred to as *lower-order interactions*.

Interarrival Distribution—In queuing theory, the distribution of the time lapses between successive arrivals at a service "counter." The exponential distribution is often used as a model for inter-arrival distributions. *See also* Arrival Distribution

Internal Data—Data taken by an organization from its own private records (as by a business firm from its stock status reports, order book, personnel sheets, etc.) for its own use in statistical studies.

Interpolation—The process of determining a value of a function between two known values without using the equation of the function itself. *See also* Extrapolation

Interquartile Range—A measure of variation given by the difference between the values of the third and first quartiles of a set of data; it represents the length of the interval which contains the middle 50 per cent of the data. A.12. *See also* Quartile Deviation

Intersection (of Two Sets)—The intersection of two sets A and B, denoted $A \cap B$, is the set which consists of all elements that belong to *both* A and B.

Interval Estimate—*See* Interval Estimation

Interval Estimation—The estimation of a parameter in terms of an interval, called an *interval estimate*, for which one can assert with a given probability (or degree of confidence) that it contains the actual value of the parameter. Note that in connection with confidence intervals the endpoints of the intervals are the random variables, but in the theory of fiducial intervals the probability applies to the parameter. *See also* Confidence Interval; Confidence Set (or Region); Fiducial Limits

Interval Scaling—*See* Scaling

Inventory Ordering Policy—A policy used in controlling an inventory system which specifies how much to order and when to order. Also called an "operating doctrine," where possible it is usually chosen so as to minimize expected cost or maximize expected profit. Under certain restrictive conditions, an (s,S) policy, for example, is one such optimal ordering policy. *See also* (s,S) Policy

Inventory System Problems—Problems arising from the needs of producers, manufacturers, sellers, etc., to carry and control stocks of raw materials, goods in process, or finished products. Basic to all such problems are the decisions of how much to order and when to order, both of which are usually complicated by the probabilistic nature of (1) the demand for goods, and (2) the interval between the time a replenishment order is placed and the time when the goods are in stock and ready to use (called the "lead time").

Inverse Matrix—*See* Matrix

Irregular Variation—In time series analysis, a term used to describe all fluctuations other than those systematic ones attributed to trend, seasonal, and cyclical influences. Thus, irregular variations are the ever-present, more or less random, movements which, though individually unpredictable, tend to average out in the long run.

Item (of a Test)—A single question, problem, or exercise on a test.

Item Analysis—In educational and psychological statistics, a procedure used to evaluate the quality of single test items and to obtain information on patterns of individual responses. Item analysis usually includes computation of an index of difficulty value for each item and an index of the extent to which each item differentiates among students scoring high and low according to chosen criteria. *See also* Difficulty Value (of an Item)

Iterative Procedure—A method of successive approximation, also called an *iteration*, where each step is based on results obtained in the preceding step (or steps).

J

***J*-Shaped Distribution**—A frequency distribution having the general shape of a letter *J* (lying on its side).

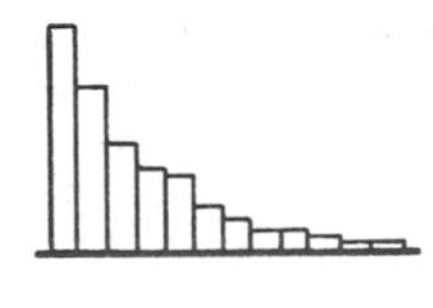

J-Shaped Distribution

Joint Density Function—An extension of the concept of a probability density function to two or more continuous random variables. **E.16.** Joint density functions are also referred to as joint probability densities and as *multivariate density functions.* In the case of two continuous random variables, probabilities are given by appropriate volumes under the surface representing their *bivariate density function.*

Joint Distribution—An extension of the concept of the distribution of a random variable to that of two or more random variables. Joint distributions are also referred to as *multivariate distributions,* and in the case of two random variables, as *bivariate distributions.*

Joint Distribution Function—An extension of the concept of a distribution function to two or more random variables. **E.17.** Joint distribution functions are also referred to as *multivariate distribution functions.* In the case of two random variables X_1 and X_2, the values $F(t_1, t_2)$ of their *bivariate distribution function* are the probabilities that X_1 assumes a value less than or equal to t_1 while, at the same time, X_2 assumes a value less than or equal to t_2.

Joint Probability Function—An extension of the concept of a probability function to two or more discrete random variables. **E.15.** Joint probability functions are also referred to as *multivariate probability functions,* and in the case of two random variables, as *bivariate probability functions.*

Judgment Sample—Unlike a probability sample, a sample in whose selection personal judgment plays a significant part. Though judgment samples are sometimes required by practical considerations, and may lead to satisfactory results, they do not lend themselves to analysis by standard statistical methods.

K

Kendall's Tau—A measure of rank correlation based on the number of inversions (interchanges of rank) required to make one ranking agree with another. **I.30.**

Kolmogorov-Smirnov Tests—Tests for significant differences between two cumulative distributions. **H.11** and **H.12.** The one-sample test is a test of *goodness of fit,* and it concerns the agreement between an observed cumulative distribution and an assumed distribution function of a continuous random variable; the two-sample test concerns the agreement between two observed cumulative distributions.

Kruskal-Wallis Test—Also referred to as the *H test,* a rank-sum test of the null hypothesis that k independent random samples come from identical populations; it is a nonparametric alternative for a one-way analysis of variance. H.6.

Kurtosis—The relative peakedness or flatness of a distribution; it is usually measured by the statistic α_4 (*alpha-four*), whose value is equal to three for the normal distribution. A.19 and E.14. A distribution which is more peaked and has relatively wider tails than the normal distribution is said to be *leptokurtic*, and for such a distribution α_4 exceeds three. A distribution which is less peaked and has relatively narrower tails than the normal distribution is said to be *platykurtic*, and for such a distribution α_4 is less than three.

L

Lag Correlation—An extension of the concept of serial correlation to two time series. It is a function of the time lag τ, that is, it measures the correlation between the values of one series at times t_1, t_2, t_3, . . . , and those of the other series at times $t_1 + \tau$, $t_2 + \tau$, $t_3 + \tau$, . . . Lag correlations are also referred to as *cross correlations*, and they are defined similarly for continuous time series.

Laspeyres' Index—A weighted aggregative index in which the prices are weighted by the corresponding quantities produced (consumed, sold, etc.) in the base year. K.9.

Latin Square—(1) In mathematics, a square array of Roman letters where each letter appears once and only once in each row and in each column. (2) In statistics, an experimental design in which the Roman letters represent different treatments. Such a design permits the simultaneous study of three sources of variation: variables represented, respectively, by the rows; the columns; and the Roman letters. The following is an example of a 4 by 4 Latin square

$$\begin{array}{cccc} A & B & C & D \\ B & C & D & A \\ C & D & A & B \\ D & A & B & C \end{array}$$

See also Greco-Latin Square

Lattice Design—A lattice is an incomplete block design suitable for handling k^2 treatments with relatively few replications of k blocks of k units; a *rectangular lattice* is suitable for handling $k(k + 1)$ treatments in an appropriate number of replicates of $k + 1$ blocks of k units. The advantage of lattice designs is that a large number of treatments can

Lattice Design – (continued)
be handled in small incomplete blocks without requiring extensive replication.

Law of Large Numbers – Informally, the *weak* law of large numbers states that if an experiment is repeated again and again, one can assert with a probability close to 1 that the proportion of the time a given event occurs will come arbitrarily close to the probability of the event (on an individual trial); the *strong* law of large numbers states that for an infinite number of trials the corresponding probability is actually equal to 1. Formal definitions may be found in the book by W. Feller listed on page 194.

Least Squares Estimate – An estimate obtained by the method of least squares; such estimates are used extensively in regression analysis and in analysis of variance. *See also* Gauss-Markov Theorem

Least Squares, Method of – A method of curve fitting and, hence, a method of estimating the parameters appearing in the corresponding equations. It consists of minimizing the sum of the squares of the differences between observed values and the corresponding values calculated by means of a model equation. For example, when fitting a straight line $y = a + bx$ to a set of paired data by this method, one minimizes (with respect to a and b) the sum of the squares of the differences between the observed y's and the corresponding values obtained by substituting the given x's into the expression $a + bx$.

Leontief Model – *See* Input-Output Analysis

Leptokurtic Distribution – *See* Kurtosis

Level (of a Factor) – In factorial experimentation, the values at which a factor (variable) is held fixed in an experiment are referred to as its levels. For example, if certain measurements are made at temperatures of 120°, 150°, and 180°, these three values are referred to as the three levels of temperature; similarly, if persons are classified as employed and unemployed in a certain study, these are the two levels of employment considered in the investigation.

Level of Significance – In a significance test, the probability of erroneously rejecting a true null hypothesis, namely, the probability α of committing a Type I error.

Library of Subroutines – A collection of subroutines for performing various standard numerical operations (computing a square root, for example) which are frequently required in automatic computing. The availability of such a collection ("library") of tested subroutines greatly expands the power and efficiency of a computer facility.

Life Testing – In engineering, the term refers to tests performed in order to determine statistical characteristics concerning the length of time a product will function, such as the *mean time to failure*, or the *MTBF*

Life Testing—(continued)

(*mean time between failures*) in processes where items (units, components, etc.) are replaced with new ones the moment they fail.

Likelihood Function—The probability or probability density of obtaining a given set of sample values from a certain population, when this probability or probability density is looked upon as a function of the parameter (or parameters) of the population and not as a function of the sample data. *See also* Maximum Likelihood Method

Likelihood Ratio Test—A test based on the ratio of the values of two likelihood functions, one of which is maximized with respect to the values of the parameter assumed under the null hypothesis, while the other is maximized with respect to the values of the parameter assumed under the alternative hypothesis (or without restriction). The ratio itself is referred to as a *likelihood ratio statistic.* Tests of this kind are known to have many desirable properties.

Limits of Prediction—*See* Prediction, Limits of

Linear Combination—In mathematics, a linear combination of x_1, x_2, $\ldots$, and x_k is an expression of the form $a_1x_1 + a_2x_2 + \ldots + a_kx_k$, where a_1, a_2, $\ldots$, and a_k are constants.

Linear Correlation—The relationship between two or more random variables for which the regression equations are linear.

Linear Discriminant Function—A discriminant function whose values are linear combinations of the values of selected variables. *See also* Discriminant Function

Linear Equation—An equation of the form $y = b_0 + b_1x_1 + b_2x_2 + \ldots + b_kx_k$, where y is the dependent variable, x_1, x_2, $\ldots$, and x_k are the independent variables, and b_0, b_1, $\ldots$, and b_k are constants. Such equations can be written in various alternate (though equivalent) forms. When a common solution is to be obtained for several linear equations, one refers to these equations as a system of *simultaneous linear equations*. *See also* Determinant

Linear Estimate—An estimate which is given by a linear combination of observed values of random variables. Examples of linear estimates are those of the regression coefficients when fitting a straight line by the method of least squares; the formulas for a and b in (I.1c) and (I.1d) can be written as linear combinations of the y's.

Linear Hypothesis—*See* Linear Model

Linear Model—A mathematical model in which the equations relating the random variables and parameters are linear. In analysis of variance, for example, where the assumption of this kind of model is referred to as a *linear hypothesis*, it is assumed that an observed value is given by the sum of terms representing different effects (main effects, treatment

Linear Model – (continued)
effects, block effects, interactions, etc.) and a value of a random variable. (J.1a) and (J.3a).

Linear Programming – The mathematical theory of maximizing (or minimizing) a linear function of several variables, when some or all of the variables are subject to restrictions (constraints) expressed in terms of linear equations or linear inequalities. Linear programming problems arise in many situations concerning the allocation of scarce resources, problems of production and consumption, the study of economic equilibrium, and also in the solution of nontrivial zero-sum two-person games. A general introduction to this subject may be found in the book by W. W. Garvin listed on page 194.

Linear Regression – *See* Regression

Linear Trend – In time series analysis, a secular trend which is represented (that is, reasonably well fit) by a straight line. The constants appearing in the trend equation are usually estimated by the method of least squares. I.1.

Link Relatives Method – A once widely used, but nowadays largely neglected, method of determining seasonal indexes.

Location, Measures of – Statistical descriptions, such as the mean, median, or a quartile, which have the following property: if the same constant is added to each observation, this constant must also be added to the measure of location.

Location Parameter – A parameter of a distribution which has the following property: if a constant is added to each value of a random variable having the given distribution, then the same constant must be added to the parameter. For example, a population mean is a location parameter, whereas a population standard deviation is not. *See also* Scale Parameter

Logarithmic Line Chart – A graph obtained by plotting the values of a series on semilogarithmic paper and connecting successive points by means of straight lines.

Logic (of a Digital Computer) – The manner in which the various logic devices are interconnected in the arithmetic-logic and control sections (units) of a computer, and how these sections are connected to the input-output and storage sections of the machine. A diagram showing in symbolic form the whole, or parts of the whole, logic design is called a *logic diagram.*

Logistic Curve – A growth curve given by an equation of the form $1/y = c + ab^x$; it may also be described as a *modified exponential curve* relating $1/y$ to x. I.6. The over-all shape of this curve (that of an elongated letter S) makes it suitable to describe the growth of human populations and that of economic variables.

Logit—In dosage problems (bio-assay), a transformed value of the probability of obtaining a particular response to a given dosage. Further information about this kind of transformation may be found in the book by D. J. Finney listed on page 194.

Log-Normal Distribution—If the logarithms of the values of a random variable have a normal distribution, the random variable itself is said to have the log-normal distribution. Some properties of the log-normal distribution are discussed in the book by I. Miller and J. E. Freund listed on page 194.

Long-Term Trend—*See* Secular Trend

Lorenz Curve—In economics, a curve designed originally to facilitate studies of the concentration of income and wealth, but having rather more general usefulness. It is obtained by plotting the cumulative percentage distribution of the variable under consideration against the cumulative percentage distribution of the number of individuals posessing given amounts of the variable. With such a curve it is easy to determine, for example, what percentage of all wage earners receive any given percentage of the total wages.

Loss Function—*See* Decision Theory

Lot—In sampling inspection, a number of relatively homogeneous units which (usually) have been collected or grouped together for inspection purposes; for instance, a set of items produced by a machine in an hour of operation.

Lot-by-Lot Sampling Inspection—A sampling plan whose purpose is to control the acceptance of "good" lots (those containing at most an allowable fraction of defective items) and the rejection of "bad" lots. Such a plan consists of drawing a sample, or samples, from each lot and determining its disposition according to the observed quality of the sample.

Lot Tolerance Per Cent Defective, LTPD—The quality (per cent defective) of a lot regarded as sufficiently bad so that it is to be accepted only with a low probability, usually 0.10. The probability that a lot of LTPD quality is accepted is called the "consumer's risk."

Lower Control Limit, LCL—*See* Control Chart

Lower-Order Interaction—*See* Interaction

M

Machine Language—The language "understood" by a particular computer. The IBM 1620, for example, cannot understand the English language instruction "Add A to B and replace A by their sum." In its language, however, the instruction 211500016000, when executed, causes precisely this operation to be performed (assuming that 15000 and 16000

Machine Language – (continued)
are the addresses at which the numbers A and B are located in the storage section of the machine); 21 is the machine language equivalent of "add."

Macro-Instruction – A single instruction (for example, "Read a Tape") in a computer source program which, during translation, generates more (often many more) than one object program instruction.

Main Effect – In factorial experimentation, the *average* change in response produced by changing the level of *one* factor; the changes in response are averaged over all possible combinations of the levels of the other factors.

Mann-Whitney Test – A rank-sum test, also called the *U test* and the *Wilcoxon two-sample test*, of the null hypothesis that two independent samples come from identical populations. H.5. It is a nonparametric alternative for the two-sample t test.

Marginal Distribution – The distribution of a random variable (or a set of random variables) obtained from a joint distribution by summing out, or integrating out, all the other variables. E.18.

Markov Chain – A discrete random process in which the probability that a system will be in a given state on the $(k + 1)$ st trial depends only on the state the system is in on the kth trial. When the possible number of states is *finite*, the process is referred to as a *finite Markov chain*. The probability p_{ij} that the system goes in one step from state E_i to state E_j is referred to as a *transition probability;* correspondingly, the probability $p_{ij}^{(n)}$ that the system does so in n steps is referred to as a *higher-transition probability*. The matrix whose elements are the transition probabilities p_{ij} is called a *transition matrix;* it is a *stochastic* matrix since all the elements are non-negative and the sum of the elements in each row is 1.

Martingale – A random process, or sequence of random variables, which is such that the conditional expectation of the $(n + 1)$ st random variable (for given values of the first n random variables) equals the value of the nth random variable. Originally, the term was used in connection with betting schemes in which losses are recovered (in theory, at least) by progressively increasing the stakes.

Master Sample – A sample drawn for repeated use, perhaps for subsequent subsampling. Such samples are used, for example, in obtaining periodic ratings of television programs.

Matched Pairs – *See* t Test; Sign Tests; Wilcoxon Test

Matched-Pairs Signed-Rank Test – *See* Wilcoxon Test

Mathematical Expectation – The mathematical expectation of a random variable, or simply its *expected value*, is given by the mean of its distribution. Originally, the concept of a mathematical expectation

Mathematical Expectation—(continued) was introduced with reference to games of chance where, if a player stood to win an amount a with the probability p, his mathematical expectation was defined as the product $a \cdot p$. A generalization of this definition is given in D.13.

Mathematical Programming Methods—A collection of techniques devised primarily to solve problems requiring that some function of a number of variables be maximized or minimized subject to a number of restrictions (constraints). *See also* Dynamic Programming; Integer Programming; Linear Programming; Nonlinear Programming

Matrix—An m by n matrix is a rectangular array having m rows, each containing n numbers, called its *elements*. If the element of the ith row and jth column is denoted a_{ij}, the corresponding 3 by 3 matrix is

$$\begin{pmatrix} a_{11} & a_{12} & a_{13} \\ a_{21} & a_{22} & a_{23} \\ a_{31} & a_{32} & a_{33} \end{pmatrix}$$

A matrix such as the above which has $m = n$ is called a *square matrix* of *order n*. A matrix A^{-1} is called the *inverse* of the square matrix A (with elements a_{ij}) if the product of the two matrices is a matrix, called the *identity matrix* and denoted by I, having 1 s on the main diagonal running from upper left to lower right and 0 s elsewhere.

Maximin Criterion—*See* Minimax Criterion.

Maximum Likelihood Estimate—*See* Maximum Likelihood Method

Maximum Likelihood Method—A method of estimation in which a parameter is estimated by the value (of the parameter) which maximizes the likelihood function. The method can also be used for the simultaneous estimation of several parameters. Estimates obtained by this method are referred to as *maximum likelihood estimates*.

McNemar Test—A nonparametric test applicable to the analysis of "correlated proportions." It may be used, for example, to analyze data pertaining to voters' preferences concerning two candidates before and after listening to a political debate. A discussion of this test may be found in the book by S. Siegel listed on page 195.

Mean—(1) The mean of n numbers is their sum divided by n. A.1 and B.1. Technically referred to as the *arithmetic mean* and commonly referred to as the "average," the mean is by far the most widely used measure of the middle, or center, of a set of data. It is usually denoted by symbols such as $\bar{x}$ or $\bar{y}$, depending on whether the observations themselves are represented by the letters x or y. (2) The mean of a finite population of size N is given by the sum of its elements divided by N. C.1. Population means are usually denoted by the Greek letter μ (*mu*).

Mean – (continued)

The mean of the distribution of a random variable is its *expected value,* see E.5 and E.4a, and it is usually denoted by μ. *See also* Geometric Mean; Harmonic Mean; Modified Mean; Weighted Mean

Mean Deviation – A measure of the variation of a set of data, also called the *average deviation,* which is given by the (arithmetic) mean of the absolute deviations from the mean. A.9 and B.7.

Mean Square – In analysis of variance, a sum of squares divided by the corresponding number of degrees of freedom. *See also* Error Mean Square; Expected Mean Square

Mean Square Error – A measure of the error to which one is exposed by the use of an estimator, which is given by the expected value of the squared difference between the estimator and a (theoretically) correct value. For unbiased estimators, the mean square error equals the variance. The square root of the mean square error is referred to as the *root-mean-square error.*

Measure of Location – *See* Location, Measures of

Measure of Variation – *See* Variation, Measures of

Median – (1) For ungrouped data, the value of the middle item (or, by convention, the mean of the values of the two middle items) when the items in a set are arranged according to size. A.5. (2) For a frequency distribution, the number corresponding to the point of the horizontal scale through which a vertical line divides the total area of the histogram of the distribution into two equal parts. B.3. (3) For the distribution of a random variable, the value (or any one of the set of values) for which the distribution function equals 1/2, or a point of discontinuity, say x_0, such that the value of the distribution function is less than 1/2 for $x < x_0$ and greater than 1/2 for $x > x_0$.

Median Test – A nonparametric test of the null hypothesis that k independent random samples come from identical populations. H.4.

Memory Section (of a Digital Computer) – *See* Digital Computer

Method of Least Squares – *See* Least Squares, Method of

Method of Semi-Averages – *See* Semi-Averages, Method of

Method of Simple Averages – *See* Simple Averages, Method of

Mid-Quartile – The mean of the values of the first and third quartiles; hence, an estimate of the middle (or mean) of a symmetrical distribution. For estimating the mean of a normal population it is relatively more efficient than the median, but less efficient than the mean.

Mid-Range – The mean of the smallest and largest values in a sample. For instance, the mid-range of the weights 110, 102, 105, 98, and 99 pounds is $\frac{110+98}{2} = 104$ pounds. The mid-range is sometimes used as a "quick" estimate of the mean of a symmetrical population; its

Mid-Range – (continued)
efficiency relative to the sample mean decreases as the sample size is increased.

Military Standard 105D (MIL-ST-105D) – A set of acceptance sampling plans, adopted by the U.S. Department of Defense, which is widely used by the government and private industry to determine the disposition (acceptance or rejection) of literally billions of dollars worth of materials and products of all kinds procured throughout the economy each year.

Minimal Complete Class (of Decision Functions) – *See* Complete Class of Decision Functions

Minimax Criterion – In decision theory and game theory, a decision criterion based on maximizing losses or expected losses with respect to the variable which one cannot control (for example, one's opponent's choice of strategies), and then minimizing with respect to the variable which one can control (for example, one's own choice of strategies). If, instead, one minimizes winnings or expected winnings with respect to the variable which one cannot control, and then maximizes with respect to the variable which one can control, the criterion is referred to as the *maximin criterion.* Estimators obtained by applying the minimax criterion to a suitable risk function are called *minimax estimators*, and strategies chosen by applying the minimax criterion to a payoff matrix are called *minimax strategies.*

Minimax Estimator – *See* Minimax Criterion

Minimax Regret Principle – In game theory, the principle of choosing the strategy which minimizes the maximum regret. For any pair of strategies of the two players in a rectangular game, the *regret* (that is, the entry of the *regret matrix*) is obtained by subtracting the corresponding entry of the payoff matrix from the maximum value of its column.

Minimax Strategy – *See* Minimax Criterion

Minimum-Variance Estimator – An estimator having the smallest possible variance within a given class of estimators, usually unbiased estimators. *See also* Cramer-Rao Inequality; Gauss-Markov Theorem

Mixed Model – In analysis of variance, the model used when some, but not all, of the parameters of the fixed-effects model are themselves values of random variables. *See also* Fixed-Effects Model; Random-Effects Model

Mixed Sampling –Multi-stage sampling, where different sampling procedures are used in the different stages; for instance, when cluster sampling is used first to select certain geographic regions and simple random samples are then selected from each of these regions.

Mixed Strategy – *See* Strategy

Modal Class – That class of a frequency distribution which has the highest frequency; sometimes a class of a frequency distribution which has a higher frequency than both adjacent classes.

Mode – (1) A measure of location defined simply as the value (or, in the case of qualitative data, the attribute) which occurs with the highest frequency, namely, most often. Note that a set of data (or a distribution) can have more than one mode, or no mode at all when no two values are alike. (2) For the distribution of a random variable, a mode is a value of the random variable for which the probability function or the probability density has a relative maximum. *See also* Multi-Modal Distribution

Model – A theory, usually expressed mathematically, which attempts to describe the inherent structure of selected aspects of a phenomenon, or process, which generates observed data. An equation which expresses a relationship among pertinent variables of a model is referred to as a *model equation*. *See also* Linear Model; Multiplicative Model

Model Equation – *See* Model

Modified Exponential Curve – A growth curve which is given by an equation of the form $y = c + ab^x$. **I.5.**

Modified Mean – When one or more of the items (usually lying at the extremes of the data) are omitted because they are judged to be atypical of the data, the mean of the remaining values is referred to as a modified mean.

Moment Generating Function – A function associated with a random variable X, whose values $M(\theta)$ equal the expected value of $e^{\theta X}$; the coefficient of $\theta^k/k!$ in its series expansion equals the kth moment of X about the origin. **E.9.**

Moments – (1) The kth moment *about the origin* of a set of data is the mean of the kth powers of the observations; the kth moment *about the mean* is the mean of the kth powers of their deviations from the mean. **A.17** and **B.14.** Moments are used to describe sets of data; for example, the mean is the first moment about the origin, and the variance is $\frac{n}{n-1}$ times the second moment about the mean. (2) For the distribution of a random variable X, the kth moment *about the origin* is the expected value of X^k, and the kth moment *about the mean* is the expected value of $(X - \mu)^k$. **E.8.**

Moments, Method of – A method of estimating the parameters of a distribution by equating as many moments of the observed data as necessary to the corresponding moments of the distribution, and solving the resulting equations.

Monte Carlo Methods – Methods of approximating solutions of problems in mathematics (and related problems in the natural and social

Monte Carlo Methods—(continued)
sciences) by sampling from simulated random processes. Such sampling is usually performed with the use of random numbers (or random-noise generators in the continuous case) and special computer techniques.

Monthly Trend Increment—The typical month-to-month change in monthly data (for example, the average monthly increase in monthly sales). It is usually obtained by converting the corresponding annual trend increment for annual data to a form reflecting the monthly increment for monthly data.

Most Powerful Test—A test of a simple hypothesis against a simple alternative is said to be most powerful if, for a fixed probability of a Type I error, the probability of a Type II error is a minimum among all tests based on the same sample size. *See also* Power; Power Function; Uniformly Most Powerful Test

Moving Average—An artificially constructed time series in which each actual value in a series is replaced by the mean of itself and some of the values immediately preceding it and directly following it. Such series are often used to describe the "general sweep" of the development of a time series when a mathematical equation is not wanted; they also form the basis for the most widely used method of constructing a seasonal index. *See also* Smoothing

Moving Totals—In time series analysis, moving totals (from which moving averages are calculated) are figures which for any given year consist of the sum of that year's figures and those of some of the years immediately preceding it and directly following it.

Mu, μ—*See* Mean

Multicollinearity—A term used mainly in econometrics to describe a common situation in multiple regression analysis of economic data where there is such a high degree of correlation between two or more explanatory (independent) variables that it is impossible to measure accurately their individual effects on the explained (dependent) variable. For instance, if income and prices are used to explain demand, these two explanatory variables are often highly correlated.

Multinomial Distribution—An extension of the binomial distribution which applies when each trial permits k possible outcomes, the trials are independent, and the probability for each possible outcome remains constant from trial to trial. **E.29.**

Multi-Modal Distribution—A distribution with several modes, that is, with several relative maxima; such distributions often result from the mixing of several non-homogeneous sets of data.

Multi-Phase Sampling—A kind of sampling in which certain items of information are collected from all the units in a sample, then other items of information (often more detailed) are collected from a subsample. Sometimes, further phases are added beyond this kind of

Multi-Phase Sampling—(continued)
two-phase process. Multi-phase sampling differs from multi-stage sampling in that the *same* sampling units are used throughout in the former, while the elements of the subsample(s) are of an inherently different nature in the latter.

Multiple Comparisons Tests—Tests designed to show which mean (or set of means) differs significantly from which other mean (or set of means). They are used, for example, as a follow-up to F tests in analysis of variance which have yielded significant results. Among the more widely used tests of this kind are the Newman-Keuls test, the Duncan multiple range test, and procedures due to H. Scheffe and J. W. Tukey; they are discussed in the book by W. T. Federer listed on page 194.

Multiple Correlation Coefficient—A measure of the closeness of the fit of a regression plane (or hyperplane) and, hence, an indication of how well one variable can be predicted in terms of a linear combination of the others. I.34. The multiple correlation coefficient is given by the *maximum* correlation coefficient between the dependent variable and any linear combination of the independent variables.

Multiple Linear Regression—A linear regression involving two or more independent variables.

Multiple Sampling—A sampling process in which one first takes a sample of size n_1; if no decision can be reached, one then takes a sample of size n_2; if still no decision can be reached on the basis of the first two samples (combined), one then takes a sample of size n_3; and so on. If a decision must be reached on the basis of the first or second sample, the process is referred to as *double sampling;* if $n_1 = n_2 = \ldots = 1$, the process is referred to as *sequential sampling. See also* Acceptance Number

Multiple Stratification—This term is synonymous with "cross stratification," namely, stratification of a population with respect to two or more variables.

Multiplicative Model—In time series analysis, the model which assumes that each value of a series is the *product* of factors that can be attributed to the individual components (secular trend, seasonal variation, cyclical variation, and irregular variation).

Multiplicative Process—*See* Branching Process

Multi-Stage Sampling—In this kind of sampling, a population is divided into a number of primary (first-stage) units, which are sampled in some appropriate way; then the selected primary units are subdivided into smaller secondary (second-stage) units and a sample is taken from these units; further stages may follow. For example, one might first select from a given population a random sample of n_1 counties (the

Multi-Stage Sampling – (continued)
primary units), then randomly select a subsample of n_2 townships (the secondary units) from each of these counties.

Multivariate Analysis – The analysis of data consisting of n-tuples (pairs, triples, etc.) of observations or, in other words, values of random vectors. It included regression and correlation analysis, analysis of variance and covariance, linear discriminant analysis, and other techniques.

Multivariate Density Function – *See* Joint Density Function

Multivariate Distribution – *See* Joint Distribution

Multivariate Distribution Function – *See* Joint Distribution Function

Multivariate Probability Function – *See* Joint Probability Function

Mutually Exclusive Events – In probability theory, two events are mutually exclusive if, and only if, they are represented by *disjoint* subsets of the sample space, namely, by subsets which have no elements in common. An alternative definition is that two events are mutually exclusive if, and only if, their intersection has a zero probability.

N

National Income and Product Account – An account, kept by the Federal government, which presents the output of the United States both in terms of final product flows (sales to consumers, sales to investors and inventory change, sales to government and net sales to foreigners) and in terms of the basic income types generated in its production (employees' compensation, rental income, corporate profits, net interest, etc.). This account provides the starting point for the input-output accounts. *See also* Input-Output Accounts

Negative Binomial Distribution – For a series of Bernoulli trials, the negative binomial distribution gives the probability that the kth success occurs on the xth trial for $x = k, k+1, k+2, \ldots$ E.27. It is also referred to as the *Pascal distribution.*

Negative Correlation – Two variables are said to be negatively correlated when the larger values of either variable tend to go with the smaller values of the other variable, and their correlation coefficient is, in fact, negative.

Negative Exponential Distribution – *See* Exponential Distribution

Negative Skewness – *See* Skewness

Nested Experiment – Also called a *hierarchial experiment,* an experiment in which a different set of levels of a second factor is used in conjunction with each level of a first factor. Such a situation would arise, for example, if one wanted to compare the properties of single-

Nested Experiment—(continued)

engine planes and two-engine planes (the two levels of the first factor), using single-engine planes made by manufacturers A, B, and C, and two-engine planes made by manufacturers B, C, and D.

Newman-Keuls Test—*See* Multiple Comparisons Tests

Neyman-Pearson Theory—A general theory of hypothesis testing based on the concepts of Type I and Type II errors, and in particular on the concept of a power function. Detailed information about this theory may be found in the book by A. Wald listed on page 195.

Noise—In information theory, the term refers to a random disturbance which is superimposed on a signal. The term *random noise* is often used in connection with artificially generated continuous random processes.

Nominal Scaling—*See* Scaling

Noncentral Chi-Square Distribution—A distribution which is used mainly to study the power of chi-square tests. Given a random sample of size n from a normal population with the mean μ and the variance σ^2, the statistic

$$\sum_{i=1}^{n}(x_i - \mu + \delta)^2/\sigma^2$$

is a value of a random variable having the noncentral chi-square distribution for any nonzero value of the parameter δ.

Noncentral *F* Distribution—A distribution which is used mainly to study the power of F tests. If two independent random variables have, respectively, a noncentral chi-square distribution and a chi-square distribution, their ratio (multiplied by a constant) has the noncentral F distribution.

Noncentral *t* Distribution—A distribution which is used mainly to study the power of t tests. Given a random sample of size n from a normal population with the mean μ and the variance σ^2, the statistic $\sqrt{n}(\bar{x} - \mu + \delta)/s$ is a value of a random variable having the noncentral t distribution for any nonzero value of the parameter δ.

Nondetermination, Coefficient of—One minus the square of the correlation coefficient; it gives the proportion of the total variation of the dependent variable which is *not* accounted for by the linear relationship with the independent variable. **I.21.**

Nonlinear Programming—An extension of the theory of linear programming to problems of maximizing or minimizing nonlinear functions subject to general restrictions (constraints).

Nonparametric Tests—Tests which do not involve hypotheses concerning specific values of parameters. Such tests are used when the

Nonparametric Tests—(continued)
assumptions underlying so-called "standard" tests cannot be met, and they are often used as computational shortcuts. H. 1 - H.12. Many statisticians use the terms "nonparametric" and "distribution-free" interchangeably. *See also* Distribution-Free Methods and listings of individual tests

Normal Correlation Analysis—*See* Correlation Analysis

Normal Curve—The graph of a normal distribution; it has the shape of a vertical cross section of a bell. Theoretically speaking, it extends from $-\infty$ to ∞, having the horizontal axis as an asymptote.

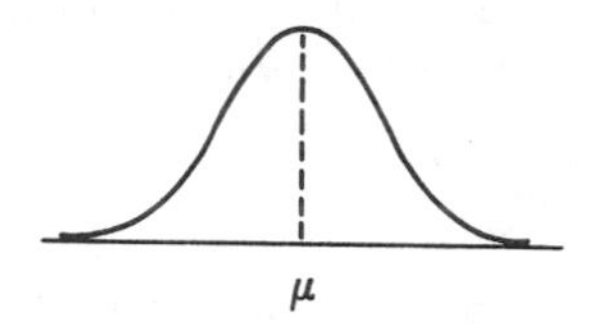

Graph of Normal Distribution

Normal Distribution—A distribution which was first studied in connection with errors of measurement and, thus, referred to as the "normal curve of errors." Nowadays, the normal distribution forms the cornerstone of a very large portion of statistical theory. Sometimes referred to also as the *Gaussian distribution,* the normal distribution has the two parameters μ and σ; when $\mu = 0$ and $\sigma = 1$ it is said to be in its *standard form,* and it is referred to as the *standard normal distribution.* E.36. Some probabilities related to the normal distribution are given on page 71.

Normal Equations—In applications of the method of least squares, a system of equations whose solution gives the least squares estimates of the parameters. For example, when fitting a straight line to a set of paired data, the normal equations are the two linear equations whose solution provides values for the regression coefficients. (I.1b), (I.4b). and (I.32b).

Normal Form (of a Game)—*See* Finite Game

Normal Inspection—In sampling inspection, the amount of inspection (sample size) used when a sampling plan is first applied. Depending on the quality of the product inspected, the inspection may subsequently be kept normal or it may be *tightened* or *reduced* (that is, the acceptance number may be lowered or increased). The effect of tightened inspection is an increase in the producer's risk and a decrease in the consumer's risk; the effect of reduced sampling is the reverse. A sampling scheme like this is used, for example, in MIL-STD-105D, a widely used plan published by the Federal government.

THE NORMAL DISTRIBUTION

The entries in this table are the probabilities that a random variable having the standard normal distribution assumes a value between 0 and z.

z	.00	.01	.02	.03	.04	.05	.06	.07	.08	.09
0.0	.0000	.0040	.0080	.0120	.0160	.0199	.0239	.0279	.0319	.0359
0.1	.0398	.0438	.0478	.0517	.0557	.0596	.0636	.0675	.0714	.0753
0.2	.0793	.0832	.0871	.0910	.0948	.0987	.1026	.1064	.1103	.1141
0.3	.1179	.1217	.1255	.1293	.1331	.1368	.1406	.1443	.1480	.1517
0.4	.1554	.1591	.1628	.1664	.1700	.1736	.1772	.1808	.1844	.1879
0.5	.1915	.1950	.1985	.2019	.2054	.2088	.2123	.2157	.2190	.2224
0.6	.2257	.2291	.2324	.2357	.2389	.2422	.2454	.2486	.2517	.2549
0.7	.2580	.2611	.2642	.2673	.2704	.2734	.2764	.2794	.2823	.2852
0.8	.2881	.2910	.2939	.2967	.2995	.3023	.3051	.3078	.3106	.3133
0.9	.3159	.3186	.3212	.3238	.3264	.3289	.3315	.3340	.3365	.3389
1.0	.3413	.3438	.3461	.3485	.3508	.3531	.3554	.3577	.3599	.3621
1.1	.3643	.3665	.3686	.3708	.3729	.3749	.3770	.3790	.3810	.3830
1.2	.3849	.3869	.3888	.3907	.3925	.3944	.3962	.3980	.3997	.4015
1.3	.4032	.4049	.4066	.4082	.4099	.4115	.4131	.4147	.4162	.4177
1.4	.4192	.4207	.4222	.4236	.4251	.4265	.4279	.4292	.4306	.4319
1.5	.4332	.4345	.4357	.4370	.4382	.4394	.4406	.4418	.4429	.4441
1.6	.4452	.4463	.4474	.4484	.4495	.4505	.4515	.4525	.4535	.4545
1.7	.4554	.4564	.4573	.4582	.4591	.4599	.4608	.4616	.4625	.4633
1.8	.4641	.4649	.4656	.4664	.4671	.4678	.4686	.4693	.4699	.4706
1.9	.4713	.4719	.4726	.4732	.4738	.4744	.4750	.4756	.4761	.4767
2.0	.4772	.4778	.4783	.4788	.4793	.4798	.4803	.4808	.4812	.4817
2.1	.4821	.4826	.4830	.4834	.4838	.4842	.4846	.4850	.4854	.4857
2.2	.4861	.4864	.4868	.4871	.4875	.4878	.4881	.4884	.4887	.4890
2.3	.4893	.4896	.4898	.4901	.4904	.4906	.4909	.4911	.4913	.4916
2.4	.4918	.4920	.4922	.4925	.4927	.4929	.4931	.4932	.4934	.4936
2.5	.4938	.4940	.4941	.4943	.4945	.4946	.4948	.4949	.4951	.4952
2.6	.4953	.4955	.4956	.4957	.4959	.4960	.4961	.4962	.4963	.4964
2.7	.4965	.4966	.4967	.4968	.4969	.4970	.4971	.4972	.4973	.4974
2.8	.4974	.4975	.4976	.4977	.4977	.4978	.4979	.4979	.4980	.4981
2.9	.4981	.4982	.4982	.4983	.4984	.4984	.4985	.4985	.4986	.4986
3.0	.4987	.4987	.4987	.4988	.4988	.4989	.4989	.4989	.4990	.4990

Normal Population – A random sample is said to come from a "normal population," if it consists of the values of independent random variables having identical normal distributions.

Normal Probability Paper – *See* Arithmetic Probability Paper

Normal Regression Analysis – *See* Regression Analysis

Normalized Standard Scores – In psychology and education, standard scores which are transformed by a special technique so that their distribution is close to a normal distribution. This kind of transformation is discussed in the second book by A. L. Edwards listed on page 193.

Norms (of a Test) – The typical or average performances of representative groups of individuals (called "norming" populations) on standardized tests. Age norms, for example, are the average values persons of various age groups can expect to score on a test. The principal types of norms used with educational and psychological tests are age, grade, percentile, and standard-score norms.

***n*-Person Game** – *See* Game of Strategy

***n*-Tuple** – In mathematics, an ordered set of n numbers; it is a generalization of such terms as "pair," "triple," etc.

Nuisance Parameter – A parameter which appears in the sampling distribution of a statistic designed to estimate *another* parameter (or test hypotheses concerning *another* parameter). For example, the population standard deviation may appear as a nuisance parameter when one estimates the mean of a population.

Null Hypothesis – Nowadays, the term is used for any hypothesis H_0 which is to be tested against an alternative hypothesis H_1 (or H_A), and whose erroneous rejection is looked upon as a Type I error. Originally, the term was used in connection with hypotheses of "no difference" in tests of significance.

Numerical Distribution – A frequency distribution, also called a *quantitative distribution,* in which data are grouped according to numerical size and not according to a qualitative description.

O

Object Program – A machine-language computer program which results from the automatic translation (as by a compiler) of a source program written in a source language, such as COBOL.

Objective Probability – In contrast to subjective probabilities, this term is applied to probabilities interpreted in the frequency sense. *See also* Frequency Interpretation of Probability

Observed Frequency—The actual number of sample values (items) falling into a class of a distribution or into a cell of a contingency table.

Ogive—The graph of a cumulative (frequency or percentage) distribution, obtained by plotting the cumulative frequencies or percentages corresponding to the class boundaries and connecting successive points with straight lines. Since most ogives met in practice have the general shape of an elongated letter S, they are also referred to as *sigmoids*.

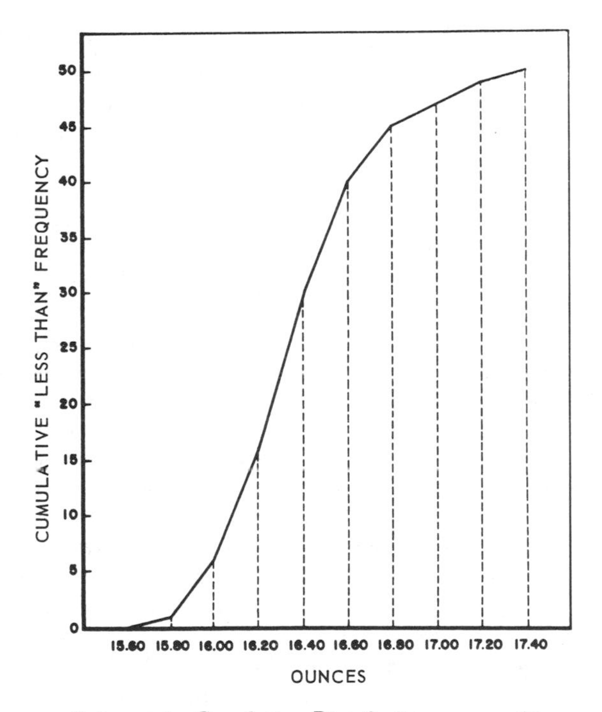

Ogive of the Cumulative Distribution on page 26

One-Sample Sign Test—*See* Sign Tests

One-Sample t Test—*See* t Tests

One-Sided Alternative—In hypothesis testing, a composite alternative hypothesis is said to be *one-sided* if the values of the parameter assumed under the alternative hypothesis are all larger or all smaller than the value (or values) assumed under the null hypothesis. Correspondingly, a composite alternative hypothesis is said to be *two-sided* if some of the values of the parameter assumed under the

One-Sided Alternative—(continued)
alternative hypothesis are larger and some are smaller than the value (or values) assumed under the null hypothesis. For example, when testing the null hypothesis $\mu = \mu_0$, the two alternative hypotheses $\mu < \mu_0$ and $\mu > \mu_0$ are both one-sided, while the alternative hypothesis $\mu \neq \mu_0$ is two-sided.

One-Tail Test (One-Sided Test)—A test of a statistical hypothesis in which the region of rejection (the critical region) consists of either the right-hand tail or the left-hand tail (but not both) of the sampling distribution of the test statistic. Correspondingly, a test is referred to as a *two-tail test* (or a *two-sided test*) when the region of rejection consists of both tails of the sampling distribution of the test statistic. For example, when testing a hypothesis concerning the mean of a population, $\bar{x} < k$ (where k is an appropriate constant) represents a one-tail test, and $\bar{x} < k_1$ or $\bar{x} > k_2$ (where k_1 and k_2 are appropriate constants) represents a two-tail test.

One-Way Analysis of Variance—An analysis of variance where the total sum of squares is expressed as the sum of the treatment sum of squares, the error sum of squares, and no others. J.1 and J.2.

One-Way Classification—A classification of a set of observations according to one characteristic (or the values of one variable). Thus, the term is used in analysis of variance when the data are grouped only according to treatments (and not according to blocks or other characteristics).

Open Class—A class at the lower or upper end of a frequency distribution having no stated lower or upper limit. For instance, classes such as "less than $100" or "$1,000 or more" are open classes.

Operating Characteristic Curve (OC-Curve)—In hypothesis testing (especially in sampling inspection), the graph of a function whose values are the probabilities of accepting the null hypothesis for various values of the parameter under consideration. Thus, for all values of the parameter other than those assumed under the null hypothesis, the OC-curve gives the probability β of committing a Type II error; for the values of the parameter assumed under the null hypothesis it gives the probability of *not* committing a Type I error (the probability $1 - \alpha$ when the parameter assumes only one value under the null hypothesis). *See also* Power Function

Operations Research—The application of modern scientific techniques to problems involving the operation of a "system" looked upon as a whole, say, the conduct of a war; the management of a firm; the manufacture of a product; the planning of an economy; and so on.

Optimum Allocation—In stratified sampling, the allocation of portions of the total sample to the individual strata so that the variance of the

Optimum Allocation—(continued)
estimator of interest is minimized. More generally, the allocation which maximizes precision for a fixed cost, or minimizes cost for fixed precision. For further information see the book by W. G. Cochran listed on page 193.

Order Statistics—The rth largest value in a sample is called the rth order statistic. Often, statistics such as the mid-range or the range, which are based on order statistics, are also referred to as order statistics.

Ordinal Scaling—*See* Scaling

Ornstein-Uhlenbeck Process—A random process which is used as a model for Brownian motion. A discussion of this process may be found in the book by E. Parzen listed on page 195.

Orthogonal Contrast—*See* Contrast

Orthogonal Latin Squares—*See* Greco-Latin Square

Orthogonal Polynomials—A set of m polynomials P_{ij} of degree i in x_j are referred to as orthogonal if for a given set of values of the x_j the sum

$$\sum_j P_{ij} P_{kj}$$

equals 0 for $i = 0, 1, \ldots, m$; $k \neq i$. Orthogonal polynomials are used extensively in fitting polynomials of degree higher than one to paired data. Illustrations of the use of orthogonal polynomials may be found in the book by R. L. Anderson and T. A. Bancroft listed on page 193.

Outliers—Observations at either extreme (small or large) of a sample which are so far removed from the main body of the data that the appropriateness of including them in the sample is questionable. Sources of outliers are gross errors in the recording of, or calculations with data, malfunctions of equipment, or contamination. *See also* Modified Mean

Output Section (of a Digital Computer)—*See* Digital Computer

P

Paasche's Index—A weighted aggregative index in which the prices are weighted by the corresponding quantities produced (consumed, sold, etc.) in the given year. **K.10.**

Paired Comparisons—Methods of establishing subjective ratings for n objects (say, n different kinds of wine, n different kinds of packaging, or n objects of art) by having judges compare them two

Paired Comparisons – (continued)
at a time and decide in each case which one is preferred. For detailed information see the book by H. A. David listed on page 193.

Paired Observations – *See* Correlated Samples; Curve Fitting; Sign Tests; t Tests

Paired-Samples Sign Test – *See* Sign Tests

Paired-Samples t Test – *See* t Tests

Parabolic Trend – In time series analysis, a trend which is best described by a parabola, that is, by a curve having an equation of the form $y = b_0 + b_1x + b_2x^2$. **I.4.**

Parameter – In statistics, a numerical quantity (such as the mean or the standard deviation) which characterizes the distribution of a random variable or a population. Parameters are usually denoted by Greek letters to distinguish them from the corresponding descriptions of samples.

Parity Ratio – An index, prepared by the U.S. Department of Agriculture, which measures whether the prices farmers receive for farm products are on the average higher or lower, in relation to the prices they pay for goods and services, than they were in the base period 1910-1914.

Partial Correlation – In multivariate problems, a measure of the strength of the relationship (correlation) between any two of the variables for fixed values of the others. **I.33.**

Partially Balanced Incomplete Block Design – An incomplete block design with all blocks of the same size, where every treatment appears the same number of times (though not more than once in each block), and where certain other conditions are satisfied that simplify the least-squares analysis. The advantage of such a design is that the number of treatments per block is small and that it avoids the large number of replications which would be required for a corresponding (completely) balanced design. Further information may be found in the book by W. T. Federer listed on page 194. *See also* Balanced Design

Partition (of a Sample Space) – The sets B_1, B_2, . . . , and B_k constitute a partition of a sample space S, if they have pairwise no elements in common and $B_1 \cup B_2 \cup \ldots \cup B_k = S$.

Pascal Distribution – An alternate name for the Negative Binomial Distribution. E.27.

Pascal's Triangle – *See* Binomial Coefficient

Payoff – *See* Payoff Matrix

Payoff Matrix – For a finite zero-sum two-person game, the payoff matrix is a matrix whose elements a_{ij} represent the *payoff*, namely the (positive, zero, or negative) amount one player receives from the other, when the first player chooses his ith pure strategy and the other player chooses his jth pure strategy.

p-Chart – In quality control, a control chart for the proportion of defective items (the fraction defective) in samples tested or inspected at regular intervals of time. L.11 and L.12.

Peakedness – *See* Kurtosis

Pearl-Reed Curve – *See* Logistic Curve

Pearson Curves – A family of distributions containing twelve general types; their density functions are all obtained as solutions of the differential equation

$$\frac{df(x)}{dx} = \frac{(x - a)f(x)}{b_0 + b_1x + b_2x^2}$$

for special values of the constants a, b_0, b_1, and b_2. A detailed discussion of Pearson curves, which include many of the distributions most commonly used in statistics, may be found in the book by M. G. Kendall and A. Stuart listed on page 194.

Pearson Product-Moment Coefficient of Correlation – *See* Correlation Coefficient

Pearsonian Coefficient of Skewness – *See* Skewness

Percentage Distribution – A converted frequency distribution in which the class frequencies are replaced by the corresponding percentages of the total number of items grouped. The following is a percentage distribution corresponding to the frequency distribution on page 44:

Weight (ounces)	*Percentage of Boxes*
15.60 - 15.79	2
15.80 - 15.99	10
16.00 - 16.19	20
16.20 - 16.39	28
16.40 - 16.59	20
16.60 - 16.79	10
16.80 - 16.99	4
17.00 - 17.19	4
17.20 - 17.39	2
	100

Percentage-of-Moving-Average Method – In time series analysis, a method of calculating a seasonal index which, until high-speed computers made possible certain refinements, was perhaps the most widely used and generally satisfactory method available. It is also referred to as the *Ratio-to-Moving-Average Method.* A discussion of this method may be found in most elementary textbooks on business statistics; for example, in the book by J. E. Freund and F. J. Williams listed on page 194.

Percentage-of-Trend Method—In time series analysis, a method of calculating a seasonal index which is generally considered to have less desirable properties than the percentage-of-moving-average method. It is also referred to as the *Ratio-to-Trend Method.* An illustration of this method may be found in the book by J. E. Freund and F. J. Williams listed on page 194.

Percentiles—The percentiles $P_1, P_2, \ldots,$ and P_{99} are values at or below which lie, respectively, the lowest 1, 2, ..., and 99 per cent of a set of data. A.8 and B.6.

Perfect Information, Game with—A game of strategy (like checkers or chess) in which the players are at all times fully informed about all previous moves.

Periodic Movement—In time series analysis, any movement which occurs more or less regularly within a prescribed period; for example, within a day, a week, a month, or a year. The main interest in the study of such phenomena has been in those movements called "seasonal variation," which recur year after year in the same months and with about the same intensity. *See* Seasonal Variation

Periodogram—In harmonic analysis, a diagram in which the intensity of a random process corresponding to a given frequency is plotted as a function of the frequency (or the corresponding wave length). For further information see the book by M. S. Bartlett listed on page 193.

Permutation—Any ordered subset of a collection of n distinct objects. The number of possible permutations, each containing r objects, that can be formed from a collection of n distinct objects is given by $n(n-1)\cdot\ldots\cdot(n-r+1) = n!/(n-r)!$, and denoted ${}_nP_r$, P_r^n, $P(n,r)$, or $(n)_r$. For instance, the number of permutations of the first ten letters of the alphabet, taken three at a time, is $10!/7! = 720$.

Personal Probability—*See* Subjective Probability

PERT (Program Evaluation Review Technique)—A management planning and control system, developed to expedite production of the Polaris missile and now widely used in the U.S. space and defense industry. Usually computer implemented, PERT (or one of its extensions) forces careful planning of the many complex and interdependent activities in multi-stage projects and, by pointing up timing and scheduling conflicts and the causes of delays, promotes on-time execution. Use of a quite similar technique, called CPM (Critical Path Method) is also widespread, especially in the construction industry.

Phi Coefficient—A measure of association (independence and dependence) in a two-by-two table, sometimes called the Yule Ø or the Yule-Boas Ø. G.17.

Pictogram—A chart used to dramatize the presentation of statistical data. Pictograms often use pictures of the objects being compared, representing differences in magnitude by repeating the picture a number of times.

TONNAGE OF VESSELS by COUNTRY 1962

U.S.
U.K.
Norway
Japan
Italy
Netherlands
France
Germany, W.
U.S.S.R.

2 million G. T.

A Pictogram

Pie Chart—A widely used chart for presenting categorical distributions, especially categorical percentage distributions. A pie chart consists of a circle subdivided into sectors whose sizes are proportional to the quantities or percentages they represent.

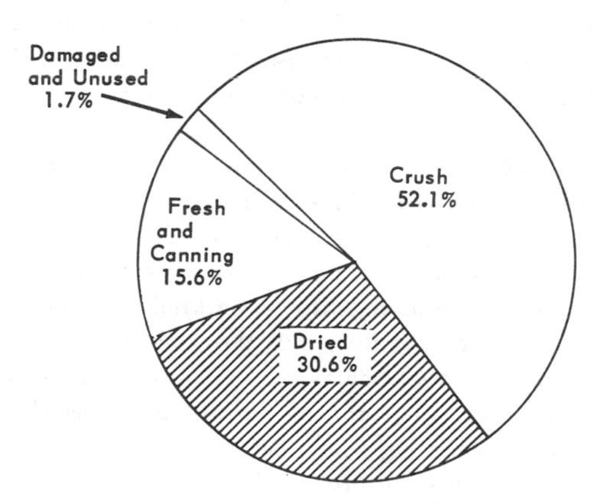

Pie Chart of the Grape Crop Utilization in California in 1963

Platykurtic Distribution—*See* Kurtosis
Plot—*See* Experimental Unit

Point Biserial Coefficient of Correlation – The correlation coefficient between a continuous variable and a dichotomous variable, with its two categories represented by the numbers 0 and 1. In contract to the biserial correlation coefficient, the point biserial coefficient of correlation is used when the dichotomous variable represents a *natural* (rather than artificial) dichotomy; that is, a dichotomy such as male and female, defective and nondefective, married and single, etc. I.27.

Point Estimate – *See* Point Estimation

Point Estimation – The estimation of a parameter by assigning it a unique value, called a *point estimate.* The merits of a method of point estimation are assessed in terms of the properties of the estimator which gives rise to the particular estimate; for example, unbiasedness, consistency, sufficiency, relative efficiency, and minimum variance.

Point of Truncation – *See* Truncated Data

Poisson Distribution – A distribution which may be introduced as a limiting form of the binomial distribution when the probability of a success on an individual trial approaches zero, the number of trials becomes infinite, and the product of these two quantities remains constant. E.28. More generally, the Poisson distribution serves as a model for situations where one is concerned with the number of "successes" per unit of observation, say, the number of imperfections per roll of cloth, the number of telephone calls arriving at a switchboard during a fixed period of time, or the daily number of automobile accidents occurring at a given intersection. Values of the Poisson probability function may be obtained from the tables by E. C. Molina listed on page 194.

Poisson Index of Dispersion – A chi-square statistic used to test whether k observed frequencies are values of independent random variables having identical Poisson distributions. G.12.

Poisson Process – A random process, continuous in time, for which the probability of the occurrence of a certain kind of event during a small time interval Δ is approximately $\alpha\Delta$, the probability of the occurrence of more than one such event during the time interval Δ is negligible, and the probability of what happens during such a small time interval does not depend on what happened before. It can be shown that under these assumptions the probability for the occurrence of x such events during a time interval of length t is given by the Poisson distribution with the parameter αt. *See also* Pure Birth Process

Polya Distribution – An alternate form of the negative binomial distribution, E.27, when k is not an integer, is referred to as the Polya Distribution. For further information see the book by W. Feller listed on page 194.

Polya Process–*See* Pure Birth Process

Polynomial Function–A function given by an equation of the form $y = b_0 + b_1x + b_2x^2 + \ldots + b_kx^k$. The highest power of x with a nonzero coefficient is referred to as the *degree* of the polynomial function. How to fit a polynomial function to a given set of paired data is explained in **I.4**.

Polynomial Trend–In time series analysis, a trend which is best described by a polynomial function. *See* Polynomial Function; Orthogonal Polynomials; Parabolic Trend

Pooled Estimate–An estimate of a parameter obtained by pooling (combining) two or more sets of data. For example, in the two-sample t test the squared deviations from the means of the two samples are pooled to obtain an estimate of the presumed common population variance.

Pooling of Error–In analysis of variance, sums of squares attributed to (supposedly nonexisting) higher-order interactions or other sources of variation are sometimes combined (pooled) with the error sum of squares to obtain more degrees of freedom for estimating the experimental error.

Population–*See* Finite Population; Infinite Population

Population Mean–*See* Mean

Population Parameter–*See* Parameter

Population Size–The number of elements in a finite population; it is usually denoted by the letter N.

Population Standard Deviation (or Variance)–*See* Standard Deviation

Positive Correlation–Two variables are said to be *positively correlated*, or have a *positive correlation*, when the larger values of either variable tend to go with the larger values of the other variable, the smaller values of either variable tend to go with the smaller values of the other variable, and their correlation coefficient is, in fact, positive.

Positive Skewness–*See* Skewness

Posterior Probability–*See* A Posteriori Probability

Postulates of Probability–*See* Probability, Postulates of

Power–For a given alternative (that is, a specific alternative value of the parameter under consideration), the probability $1 - \beta$ of not committing a Type II error with a given test. Correspondingly, one test is said to be *more powerful* than another for a given alternative if its power exceeds that of the other test. *See also* Power Function; Uniformly More Powerful Test; Uniformly Most Powerful Test

Power Efficiency–This concept concerns the increase in sample size required to make a test as powerful as the most powerful known test of its type (when used with data which meet its assumptions). It is

Power Efficiency – (continued)
given by the percentage $\frac{n_1}{n} \cdot 100$, where n_1 is the sample size required with test T_1 to make it as powerful as test T with a sample of size n. Generally, the power efficiency of a test will depend on n and also on the value of the parameter assumed under the alternative hypothesis.

Power Function – In hypothesis testing, a function whose values are the probabilities of rejecting a given null hypothesis (and accepting the alternative hypothesis) for various values of the parameter under consideration. Thus, for all values of the parameter other than those assumed under the null hypothesis the power function gives the probability $1 - \beta$ of *not* committing a Type II error; for the values of the parameter assumed under the null hypothesis it gives the probability of committing a Type I error (the probability α when the parameter assumes only one value under the null hypothesis). Observe that the values of the power function are given by 1 *minus* the corresponding values of the *operating characteristic curve.*

Precision – (1) The precision of an estimator is its tendency to have its values cluster closely about the mean of its sampling distribution; thus, it is related inversely to the variance of this sampling distribution–the smaller the variance, the greater the precision. (2) The term "precision" has been used to denote the parameter $h = 1/\sigma\sqrt{2}$ of the normal distribution.

Prediction, Limits of – In regression analysis, a pair of values for which one can assert with a given probability that they will contain a future observation of the dependent variable for a given value of the independent variable. **I.13.**

Price Index – An index number intended to estimate changes in the prices of certain goods or services; for example, wholesale food prices or consumer prices.

Price Relative – The ratio of the price of an individual item in the given year to its price in the base year. **K.1.**

Primary Data – Statistical data which are published by the same organization by which they are collected (and often processed).

Prior Density – *See* A Priori Probability

Prior Probability – *See* A Priori Probability

Probability – *See* Frequency Interpretation of Probability; Probability, Postulates of; Subjective Probability

Probability Density – This term is used as an abbreviation for "probability density function," and also to denote a *value* of such a function. *See* Probability Density Function

Probability Density Function – A function with non-negative values, whose integral from a to b ($a \leq b$) gives the probability that a corresponding random variable assumes a value on the interval from a to b. E.2. Most of the probability density functions used in basic statistics may be represented by continuous curves, and probabilities are given by appropriate areas under these curves. Probability density functions are also referred to in the literature as *continuous densities, continuous distributions, densities, density functions, frequency functions*, and *probability densities*.

Probability Distribution – (1) A synonym for "probability function." (2) The probability distribution (or simply the distribution) of a random variable is its *probability structure* as described, for example, by a probability function or a probability density function. *See also* Probability Density Function; Probability Function

Probability Function – A function which assigns a probability to each value within the range of a discrete random variable. E.1. Probability functions are also referred to in the literature as *densities, discrete density functions, discrete densities, frequency functions, probability distributions*, and *probability distribution functions.*

Probability Graph Paper – *See* Arithmetic Probability Paper

Probability Measure – A function, satisfying the postulates of probability, that is defined on the elements of a σ-field of events.

Probability, Postulates of – Basic rules, or axioms, which form the foundation of the mathematical theory of probability. D.1. These rules apply regardless of whether probabilities are interpreted as frequencies, as degrees of belief, or as logical relationships.

Probability Sample – A sample obtained by a method in which every element of a finite population has a known (not necessarily equal) chance of being included in the sample.

Probability Space – A triple which consists of a sample space S, a σ-field of events in S, and a probability measure defined on the elements of the σ-field of events.

Probable Error – A value for which one can assert with a probability of 0.50 that is will not be exceeded in size by the error of an estimate. Based on the assumption that the sampling distribution of the estimator is normal, it is given by 0.6745 times the corresponding *standard error*, namely, 0.6745 times the standard deviation of this sampling distribution. Nowadays, probable errors are used mainly in military applications.

Probit – In dosage problems (Bio-Assay), a transformed value of the probability of obtaining a certain response to a given dosage. This transformation consists of adding five (to avoid negative values) to

Probit – (continued)
the values of a random variable having the standard normal distribution, and for which the area under the curve to its left equals the probability of the response to the dosage. For details, see the book by D. J. Finney listed on page 194.

Problem-Oriented Language – A term sometimes used interchangeably with "procedure-oriented language." However, "problem-oriented language" also refers to a computer language in which only the problem is stated, the compiler thereafter choosing a procedure to solve the problem.

Procedure-Oriented Language – A computer language, FORTRAN or COBOL, for example, independent of any particular make or model of computer, in which the procedures for solving a problem may be conveniently stated.

Producer's Risk – The probability α of committing a Type I error; the term is used especially in problems of sampling inspection.

Product Moment – *See* Covariance

Program – A complete plan for solving a problem with an automatic computer. However, the term is often used to refer only to that part of the computer plan which consists of the detailed written instructions to the machine itself. *See also* Programming (a Digital Computer); Coding (a Digital Computer)

Programming (a Digital Computer) – Specifying precisely how a problem is to be solved on a digital computer. Although the terms "programming" and "coding" are often used synonymously, strictly speaking programming includes the following: problem recognition, definition, and analysis; flowcharting; coding; testing and debugging; documentation (describing the problem, the results, and the process by which they are obtained); and program maintenance. The two major types of programming are (1) *applications programming*, namely, writing programs to solve specific scientific or commercial data processing problems by utilizing the computing resources of a particular computer or computer facility. A regression analysis program written for execution on one of the IBM System/360 computers, for example, is an applications program. The other type is (2) *systems programming*, namely, writing programs which facilitate operating, and writing applications programs for, (often) a wide range of computers. Among the various products of this extremely complex and difficult work are operating systems, programming languages and compilers, and testing programs to detect mechanical and electrical equipment malfunctions.

Programming Language/One (PL/1) – A high-level computer programming language, announced in 1964, and developed for use initially with the IBM System/360 family of computers. Intended as a more or less universal language to meet most business and scientific needs, PL/1 incorporates some features of the FORTRAN, COBOL, and ALGOL languages, as well as a number of others found in none of these languages.

Proportional Allocation – In stratified sampling, the allocation of portions of the total sample to the individual strata so that the sizes of these subsamples are proportional to the sizes of the corresponding strata. For instance, if a stratified sample of 100 students is to be taken from among the 400 freshmen, 300 sophomores, 200 juniors, and 100 seniors attending an undergraduate school, proportional allocation requires that 40, 30, 20, and 10 students be chosen from these four classes.

Pseudorandom Numbers – *See* Random Numbers

Purchasing Power of the Dollar – The purchasing power (or value) of some kind of dollar (for example, a food dollar) is given by the reciprocal of an appropriate price index (for example, a food price index) written as a proportion. The purchasing power of the dollar based on the U.S. Department of Labor's Consumer Price Index and also on its Wholesale Price Index are both regularly reported by the Federal government.

Pure Birth Process – A random process representing the size of a population for which, in a very small interval of time, the population size either increases by one or stays the same. If the probability that the population is increased by one during the small time interval Δ is given by $\alpha n \Delta$ (where n is the size of the population and α is a constant), the process is referred to as a *Yule Process;* if this probability is given by $\alpha\Delta$, the process is referred to as a *Poisson Process*. If the probability that the population size is increased by one during the small time interval Δ at time t is given by $\frac{1 + an}{1 + at}\Delta$ (where n is the size of the population at time t and a is a constant), the process is referred to as a *Polya Process*.

Pure Death Process – A random process representing the size of a population, for which in a very small interval of time the population size either decreases by one or stays the same.

Pure Strategy – *See* Strategy

Q

Qualitative Distribution – *See* Categorical Distribution

Quality Control – Though sometimes viewed broadly as the application of any useful statistical techniques to the control of the quality of manufactured products and raw materials, the term refers essentially to the use of control charts and the methods of acceptance sampling. *See also* Control Chart; Acceptance Sampling

Quantile – *See* Fractile

Quantitative Distribution – *See* Frequency Distribution; Numerical Distribution

Quantity Index – An index number intended to estimate changes in the physical volume or quantity produced, consumed, sold, etc., of certain goods or services.

Quantity Relative – The ratio of the quantity produced, consumed, sold, etc., of an individual item in the given year to the corresponding quantity in the base year. **K.2.**

Quartile Deviation – A measure of variation, also called the *semi-interquartile range,* which is given by half the difference between the third and first quartiles; hence, the average amount by which the first and third quartiles differ from the median. **A.13.**

Quartile Variation, Coefficient of – A measure of relative variation in which the difference between the third and first quartiles (of a set of data or a distribution) is divided by their sum, and the quotient is multiplied by 100. **A.16.**

Quartiles – The quartiles Q_1, Q_2, and Q_3 are values at or below which lie, respectively, the lowest 25, 50, and 75 per cent of a set of data. **A.6.**

Quasirandom Numbers – *See* Random Numbers

Questionnaire – A list of questions sent (or given) to a subject in a statistical investigation which, it is hoped, he will answer completely and truthfully, and promptly return to the source.

Queue – Also called a waiting line, a line formed by the arrival of persons, items, or units seeking service of some sort at a facility which serves and subsequently releases them; for instance, trucks waiting to be unloaded at a dock, aircraft waiting to land, court cases waiting to be heard, relief cases waiting to be processed, and customers waiting to be served at a ticket office or market check-out stand all form queues. Except in special circumstances, queues eventually develop whenever units arrive at a servicing facility. *See also* Queue Discipline; Queuing Theory

Queue Discipline – The order (for example, "first come, first served") in which the units of a queue are served at a service facility. The

Queue Discipline—(continued)
discipline imposed on the waiting line affects some variables of interest (the distribution of waiting times, for instance), but it does not affect others (the distribution of queue lengths, for instance).

Queuing Theory—Sometimes called *congestion theory, trunking theory,* or *waiting-line theory,* queuing theory is a (largely) mathematical theory concerned with the study of such factors of interest in queuing problems as the distribution of arrivals, the distribution of the lengths of service times, the average time a unit spends in the system (waiting to be served and being served), the average length of the queue, and the probability distribution for the number of units waiting to be served. *See also* Arrival Distribution; Interarrival Distribution; Arrival Rate, Mean; Service Rate, Mean; Service Time Distribution

Quota Sampling—A type of judgment sampling, in which an interviewer is given a quota to be filled (for example, three American-born shopkeepers in the over-65 age bracket living in a certain area, etc.); such quotas are usually filled as quickly and simply as possible, without much consideration to questions of randomizing the selection.

R

Random-Effects Model—The model used in analysis of variance when the parameters of the fixed-effects model are, themselves, values of random variables; this model is also referred to as *Model II*, the *variance-components model*, and the *components of variance model. See also* Fixed-Effects Model; Mixed Model

Random Normal Deviates—Random numbers (usually two-decimal or three-decimal numbers) which can be regarded as values of a random variable having the standard normal distribution. Hence, they are used to simulate random sampling from normal populations.

Random Numbers—Published tables of random numbers (or *random digits*) consist of pages on which the digits 0, 1, 2, . . . , 9 are set down in much the order as they would appear if they had been generated by a gambling device which gives each digit a probability of 1/10. Before publishing tables of this sort, the numbers are usually required to "pass" various statistical tests intended (insofar as this is possible) to insure their randomness. In computer work it is generally convenient to generate sequences of random numbers as they are needed by following some calculational rule, which is devised in such a way that the numbers generated would meet various statistical tests of randomness although they are never actually tested; such numbers are called *pseudorandom numbers.* Sometimes,

Random Numbers – (continued)
it is convenient in such work to use sequences of *quasirandom numbers*, that is, "more or less" random numbers which, though failing to satisfy some statistical criterion, are nevertheless valid for use in particular situations. The RAND Corporation table of random numbers is referred to on page 195. *See also* Random Normal Deviates

Random Order – An ordering of a set of objects which is such that every possible order has the same probability. The process of arranging a set of objects in random order is referred to as *randomization;* it is used, for example, when allocating treatments to the experimental units (plots) as part of the design of an experiment.

Random Process – A random process, or *stochastic process*, is (1) a physical process which is governed at least in part by some random mechanism, and (2) a corresponding mathematical model. In the study of random processes one is generally concerned with sequences of random variables with special reference to their interdependence and limiting behavior. Examples of random processes are provided by the growth of populations such as bacterial colonies, and the fluctuating throughput in successive runs of an oil-refining mechanism. A random process can be discrete or continuous in time, and its value at any given time can be a value of a discrete or a continuous random variable. For instance, a continuous graph having the values 1 or 0, depending on whether or not a computer is in use, is continuous in time while the random variable (with values 0 and 1) observed at any given time is discrete; on the other hand if, in quality control, one takes hourly samples from normal populations and records their means, the process is discrete in time while the random variable observed at any given time is continuous. *See also* Ergodic Process; Stationary Process; and individual listings by name

Random Sample – (1) A sample of size n from a *finite* population of size N is said to be random if it is chosen so that each of the $\binom{N}{n}$ possible samples has the same probability of being selected; such samples are also referred to as *simple*, or *unrestricted*, random samples. (2) A set of observations constitutes a random sample of size n from an *infinite* population, if the n observations are values of independent random variables having the same (population) distribution.

Random Start – *See* Systematic Sample

Random Variable – Also called a *chance variable*, a *stochastic variable*, or a *variate*, a random variable is a real-valued function defined over a sample space. The number of heads obtained in five flips of a coin, the height of a person chosen for an experiment, and the number of

Random Variable – (continued)
traffic accidents in Phoenix during the month of April, are all examples of random variables. In order to distinguish between random variables (functions) and their values, it has become customary to use capital letters (as in this book) or boldface type, say X or **x**, to denote random variables, and the corresponding lower case letters, x, to denote their values. *See also* Continuous Random Variable; Discrete Random Variable

Random Vector – *See* Vector

Random Walk – A path in which each step, that is, its magnitude and/or its direction, is determined by chance. In statistics the concept arises, for example, in connection with sequential sampling.

Randomization – *See* Random Order

Randomization Tests – Nonparametric tests in which certain aspects of a sample are studied by considering all possible (equally likely) arrangements of its elements. Thus, a test of randomness based on runs and the Mann-Whitney test are, essentially, randomization tests.

Randomized Block Design – An experimental design in which the treatments in each block are assigned to the experimental units in random order.

Randomized Strategy – *See* Strategy

Randomness, Tests of – Tests which enable one to decide whether certain features of the arrangement (or order) of sample values differ significantly from what one might expect if they were drawn at random and, hence, are indicative of an element of nonrandomness. Most of these tests are randomization tests; among the most widely used are those based on the total number of runs or the total number of runs above and below the median. **H.8** and **H.9**.

Range – A simple, and easily obtained, measure of the variation of a set of data given by the difference between the largest value and the smallest. **A.14.** For instance, the range of the five lengths 3.09, 3.14, 3.07, 3.10, and 3.12 is $3.14 - 3.07 = 0.07$. As an estimator of a population standard deviation its efficiency, relative to the sample standard deviation, decreases as the sample size is increased.

Rank Correlation – The association between paired values of two random variables which have been replaced by their ranks within their respective samples (or which are originally given as ranks). *See also* Spearman's Rank Correlation Coefficient; Kendall's Tau

Rank Correlation Coefficient – *See* Spearman's Rank Correlation Coefficient; Kendall's Tau

Rank-Order Statistics – Statistics based on the rankings of observations; that is, statistics in which observations are replaced by their

Rank-Order Statistics – (continued)
ranks. Examples of rank-order statistics are the Spearman rank correlation coefficient, the U statistic used in the Mann-Whitney test, and the H statistic used in the Kruskal-Wallis test.

Ratio Estimate – In survey sampling, an estimate of a population total for one variable, which is obtained by multiplying the known population total for another variable by a ratio of appropriate sample values of the two variables. The known population total might be obtained from a census taken a number of years ago, and the corresponding current total might be estimated by multiplying the known total by a ratio (percentage change) based on subsamples. *See also* Regression Estimate

Ratio Paper – *See* Semi-Logarithmic Paper

Ratio Scaling – *See* Scaling

Ratio-to-Moving-Average Method – *See* Percentage-of-Moving-Average Method

Ratio-to-Trend Method – *See* Percentage-of-Trend Method

Rational Subgroup – In quality control, a sample of items chosen in such a way that all items are produced and inspected under as nearly identical conditions as possible; hence, a homogeneous sample in which the possibility of assignable variations within the sample (subgroup) is reduced to a minimum.

Raw Data – Data which have not been subjected to any sort of statistical treatment (such as grouping, coding, censoring, etc.).

Raw (Test) Scores – The first recorded numerical result of a test, such as the number of correct answers. Since raw scores generally are not comparable from one test to another and have no "standard" meaning, they usually are subjected to some sort of statistical treatment or analysis; for example, converted into standard scores or normalized standard scores.

Rayleigh Distribution – A term which is used mainly in engineering applications to denote a special form of the Weibull distribution. It arises as the distribution of the miss distances from a target when the coordinates of the point of impact have the circular normal distribution with its center at the target. E.40.

Real Class Limits – *See* Class Boundary

Real Income – *See* Real Wages

Real Time (Computer) Systems – Information processing systems (such as are found in process control, in many military applications, and elsewhere) in which subsequent control of the system depends entirely, or almost so, on the output of the computer. In such time-critical situations, solutions are required almost immediately after the input data are entered into the computer. When a computer

Real Time (Computer) Systems—(continued) *simulation* is performed in the same time it takes to perform the actual operation being simulated, it is said to be performed in *real time;* simulations are performed in "slow time" or "fast time" when they are performed slower or faster than the actual operations.

Real Wages—The money wages one receives after they have been adjusted for changes in the level of prices. This adjustment is accomplished by deflating the money wages (a value series) by dividing them by the values of an appropriate price index. For instance, in 1965 average weekly wages in manufacturing for a certain state were $97.55 and a relevant price index stood at 114.0 (base year 1959); thus, the real wages were (97.55/114.0)100 = $85.57. One way to express this result is to say that what could have been bought for $85.57 in 1959 cost $97.55 in 1965.

Rectangular Distribution—*See* Uniform Distribution

Rectangular Game—*See* Finite Game

Reduced Inspection—*See* Normal Inspection

Reduction of Data—*See* Data Reduction

Regression—The relationship between the (conditional) mean of a random variable and one or more independent variables; a mathematical equation expressing this kind of relationship is called a *regression equation.* When the regression equation is a linear equation the regression is also referred to as *linear;* when the regression equation represents some other kind of curve or surface the regression is referred to as *curvilinear.* I.8. The term "regression" is due to Francis Galton, who employed it first in connection with a study of the heights of fathers and sons, observing a regression (or turning back) from the heights of sons to the heights of their fathers.

Regression Analysis—The analysis of paired data (x_1, y_1), (x_2, y_2), . . . , and (x_n, y_n), where the x's are constants and the y's are values of random variables; it is referred to as *normal regression analysis* if the y's are values of independent random variables having normal distributions with the respective means $\alpha + \beta x_i$ and the common variance σ^2. The term "regression" is also applied to the analysis of n-tuples of data, where the values of the independent variables are looked upon as constants and the values of the dependent variable are values of random variables. *See also* Correlation Analysis

Regression Coefficient—A coefficient in a regression equation; for example the parameters α and β in the linear regression equation $y = \alpha + \beta x$. I.8. The term is also used for corresponding estimates, but it is preferable to refer to these specifically as *estimated regression coefficients.*

Regression Curve (or Surface) – A graphical presentation of a regression equation in two (or three) variables. Sometimes these terms are used to refer exclusively to the case where the regression equation is non-linear.

Regression Estimate – In survey sampling, an estimate of a population total for one variable which is obtained by substituting the known total for another variable into a regression equation calculated on the basis of sample values of the two variables. Note that *ratio estimates* are special kinds of regression estimates.

Regression Line (or Plane) – A graphical presentation of a linear regression equation in two or three variables.

Regret – *See* Minimax Regret Principle

Regret Matrix – *See* Minimax Regret Principle

Rejection, Region of – *See* Critical Region

Rejection Number – *See* Acceptance Number

Relative Efficiency – (1) In estimation theory, a measure comparing the *precision* of two unbiased estimators of a parameter; the efficiency of one estimator relative to a second is given by the ratio of the variance of the sampling distribution of the second to that of the first. (2) In experimental design, the ratio of the amount of replication required by two designs to attain the same precision. For further information see the book by W. G. Cochran and G. M. Cox listed on page 193.

Relative Frequency – If an event occurs x times in n trials, the relative frequency of its occurrence is x/n; relative frequencies are also referred to as *sample proportions.*

Relative Importance Weights – In the construction of the *Consumer Price Index,* the eight major groups (food, housing, etc.) are combined to form a city index by assigning each a weight intended to represent the "relative importance" of the group in the typical expenditures of the index families. Sometimes the term is used more generally for any weights employed in the construction of a weighted mean or a weighted index number.

Relative Variation – The variation of a set of measurements relative to the size of whatever object (or quantity) is being measured. The most widely used measure of relative variation is the coefficient of variation, which is given by the ratio of the standard deviation to the mean (multiplied by 100 to express the coefficient as a percentage). A.15 and A.16. For instance, if the mean and the standard deviation of a sample are 2.00 in. and 0.10 in., respectively, the coefficient of variation is $\frac{0.10}{2.00} 100 = 5$ per cent.

Reliability—(1) In statistics, the term "reliability" is often used interchangeably with "consistency;" thus, when speaking of the reliability of an estimator, one is referring to the possible size of chance errors. (2) In engineering, the reliability of a product (component, unit, etc.) is the probability that it will perform within specified limits for at least a specified length of time under given environmental conditions. (3) In educational and psychological testing, the reliability of a test is the consistency with which it measures whatever trait it does measure.

Reliability Coefficient—In psychology and education, the self-correlation of a test, or the correlation between two comparable measures of the same thing or trait. Among the schemes used for determining the reliability of a test are correlating test-retest scores, split-half scores, or scores on supposedly equivalent forms. *See also* Split-Half Method

Replacement, Sampling With and Without—If an element is drawn from a finite population, its distribution (or composition) for the next drawing is disturbed unless the element is replaced. When elements are not replaced, sampling is said to be *without replacement*, successive drawings are dependent, and the probability of getting x elements of a certain kind in n drawings is given by the hypergeometric distribution. E.26. When each element is replaced before the next one is drawn, sampling is said to be *with replacement*, successive drawings are independent, and the probability of getting x elements of a certain kind in n drawings is given by the binomial distribution. E.23. If the sample size is small compared to the size of the population (less than 5 per cent), the binomial distribution is often used as an approximation to the hypergeometric distribution when sampling is *without replacement*.

Replicate—*See* Replication

Replication—In experimental design, the performance of an experiment (or parts of an experiment) more than once; the purpose of replication is to obtain more information (more degrees of freedom) for estimating and assessing the experimental error and to obtain estimates of effects with smaller standard errors. The individual repetition of an experiment (or part of an experiment) is referred to as a *replicate*.

Representative Sample—A sample, chosen by any means, which exhibits on a small scale the relevant characteristics of the population from which it came; that is, a sample which is, in fact, representative of the population.

Residual Method—In time series analysis, a classical method of estimating cyclical components by first eliminating the trend,

Residual Method – (continued) seasonal variations, and irregular variations, thus leaving the cyclical relatives as residuals.

Residual Variance – In regression analysis and analysis of variance, that part of the variability of the dependent variable which is attributed to chance or experimental error, namely, that part of the variability of the dependent variable which is not attributed to specific sources of variation.

Response Surface Analysis – Statistical methods of prediction and optimization, including (among others) regression analysis and factorial experimentation. In particular, response surface analysis is concerned with methods leading to experimental conditions for which the response of a dependent variable (or dependent variables) is optimized (maximized or minimized, as the case may be) and with studying the response of the dependent variable in the vicinity of the point where it is optimized. *See also* Steepest Ascent, Method of

Risk, Producer's and Consumer's – In hypothesis testing, especially in acceptance sampling, the respective probabilities of committing Type I and Type II errors.

Risk Function – *See* Decision Theory

Road Map – *See* Flowchart

Robustness – A method of inference is said to be robust if it is relatively unaffected when all of its underlying assumptions are not met.

Root-Mean-Square Deviation – *See* Standard Deviation

Root-Mean-Square Error – *See* Mean Square Error

Rotatable Design – An experimental design from which equally reliable predictions can be made in all directions. That is, the reliability of a prediction depends only on the distance of an experimental condition from a central point and not on its direction from that point.

Routine – A set of coded instructions which cause a computer to perform the various operations necessary to solve a given problem; for example, a routine to evaluate a multiple integral by Monte Carlo methods.

Run – A run is a sequence of identical symbols (letters, for example), which is followed and preceded by a different symbol or no symbol at all. The total number of runs, generally denoted by the letter u, in an arrangement of two or more kinds of symbols is often used as a check on the randomness of the arrangement. **H.8**. This statistic also forms the basis of a nonparametric test of the null hypothesis that two independent random samples come from identical populations, **H.10**, and for a test of randomness based on *runs above and below the median*, when sample values are replaced by the letters

Run – (continued)
a and *b*, depending on whether they fall above or below the median of the sample. H.9.

Runs Up-and-Down – In time series analysis, if differences between successive values of a series are represented by + or −, depending on whether there is an increase or a decrease, runs in the resulting sequence of plus and minus signs are referred to as *runs up-and-down.* The total number of such runs up-and-down forms the basis of non-parametric tests of randomness against the alternative that there is a trend or, perhaps, a cyclical pattern. For further information see the book by E. S. Keeping listed on page 194.

S

Saddle Point – In a zero-sum two-person game, the pure strategies of two players constitute a saddle point if the corresponding entry of the payoff matrix is the minimum value of its row and the maximum value of its column. Games having saddle points are said to be *strictly determined;* strategies corresponding to a saddle point are optimum strategies for the two players.

Sample – *See* Judgment Sample; Probability Sample; Random Sample; Representative Sample; Sample Design

Sample Census – A detailed census, or examination, of individuals, households, businesses, etc., selected in a sample from some population.

Sample Design – A definite plan, completely specified before any data are actually collected, for obtaining a sample from a given population. Alternate terms are "sampling plan" and "survey design." *See also* Area Sampling; Cluster Sampling; Judgment Sample; Multi-Phase Sampling; Multi-Stage Sampling; Probability Sample; Quota Sampling; Random Sample; Representative Sample; Sequential Sampling; Stratified Random Sampling

Sample Mean – *See* Mean

Sample Proportion – *See* Relative Frequency

Sample Size – The number of observations in a sample; it is usually denoted by the letter *n*.

Sample Space – In probability, a set of points (elements) which represent all possible outcomes of an experiment.

Sample Standard Deviation – *See* Standard Deviation

Sample Survey – A survey of human populations, businesses, social institutions, etc., which is based on samples. *See also* Sample Census; Sample Design

Sample Variance—*See* Standard Deviation

Sampling—The process of obtaining a sample. *See also* the various kinds of samples and methods of sampling referred to under Sample Design

Sampling Distribution—The distribution of a statistic; for example, the distribution of the sample mean for random samples from normal populations, or the distribution of the coefficient of correlation under the assumptions of normal correlation analysis. *See also* Experimental Sampling Distribution

Sampling Error—(1) In general, the difference between an observed value of a statistic and the quantity it is intended to estimate. (2) In analysis of variance, the following distinction is made between *sampling error* and *experimental error:* if there are several observations per cell, namely, if several observations are made on *one* experimental unit, their variation is referred to as *sampling error;* on the other hand, differences not attributed to specific sources of variation among observations made on *different* experimental units are referred to as *experimental error.*

Sampling Inspection—*See* Acceptance Sampling; Continuous Sampling Inspection; Lot-by-Lot Sampling Inspection; Military Standard 105D (MIL-ST-105D)

Sampling Plan—*See* Sample Design

Scale Parameter—A parameter of a distribution having the following property: if each value x of a random variable having a given distribution is replaced by $ax + b$, then the parameter must be multiplied by the constant a. For example, a population standard deviation is a scale parameter, whereas a population mean is not. Generally, parameters which are given by powers of scale parameters are also referred to as scale parameters; in this sense, a population variance would be considered a scale parameter. *See also* Location Parameter

Scaling—In general, scaling is a process of measuring. In particular, *nominal scaling*, or *categorical scaling*, is the process of grouping individual observations into qualitative categories, or classes; for example, classifying hospital patients according to their disorders. *Ordinal scaling* is a measuring procedure which assigns one object a greater number, the same number, or a smaller number than a second object only if the first object posesses, respectively, more, the same amount, or less of the characteristic being measured than the second object; for example, rating the hardness of five minerals by means of scratch tests, assigning the number one to the softest (which is scratched by all and scratches no others) and the number five to the hardest (which scratches all and is scratched by no others). *Interval scaling* is a special kind of ordinal scaling where the measurement

Scaling—(continued)
assigned to an object is linearly related to its true magnitude; in other words, in interval scaling the scale of measurement has an arbitrary origin and a fixed, though arbitrary, unit of measurement (as, for example, when finding temperatures in Fahrenheit units). *Ratio scaling* is a special kind of interval scaling where the measurement assigned to an object is proportional to its true magnitude; in other words, the scale of measurement has an absolute zero and a fixed, though arbitrary, unit of measurement (as, for example, when measuring length or weight).

Scatter Diagram—A set of points obtained by plotting paired measurements, as points in a plane. The visual inspection of such diagrams aids in analyzing the general nature of the relationship between the two variables.

Schedule—A form on which an agent, enumerator, or interviewer enters the answer to questions asked directly of a subject; hence, a kind of questionnaire.

Scheffé's Test—*See* Multiple Comparisons Tests

Scientific Notation—A widely used scheme for writing numbers in scientific work. In such notation, numbers are written as a number between 1 and 10 multiplied by a power of 10. For instance, 530,000,000,000 is written $5.3 \cdot 10^{11}$ and 0.00000072 is written $7.2 \cdot 10^{-7}$. *See also* Floating Point Arithmetic

Screening—(1) In quality control, a 100 per cent inspection procedure in which each item is inspected, all good items are accepted, and all bad items are rejected. (2) In computer operations, *data screening* procedures are used to check for outliers, for example, or to detect implausible (or plausible but probably wrong) data input to the computer.

Seasonal Index—For monthly data, a set of twelve numbers (one for each month), in which each month's activity is expressed as a percentage of that of the average month. For example, a January index of 71 means that January figures are typically 71 per cent of those of the average month. Correspondingly, a seasonal index may also be defined for weekly or daily data. *See also* BLS Seasonal Factor Method (1964)

Seasonal Variation—Strictly speaking, the movements in a time series which recur year after year in the same months with more or less the same intensity. Although the name implies a connection with the seasons of the year, the term is often used somewhat loosely to indicate other periodic movements; for example, those occurring within a day, week, or month. *See also* Periodic Movement

Secondary Data—Statistical data which are published by some organization other than the one by which the data are collected (and often processed).

Secular Trend—Sometimes called a *long-term trend,* the secular trend of a time series ordinarily is the underlying smooth (or regular) movement of the series over a fairly long period of time. This underlying movement is thought to result from persistent forces slowly affecting growth or decline (changes in population, income and wealth, changes in the level of education and technology, etc.).

Semi-Averages, Method of—A simple, though very crude, method of fitting a straight line trend to a time series. This method is illustrated in the book by J. E. Freund and F. J. Willians listed on page 194.

Semi-Interquartile Range—*See* Quartile Deviation

Semi-Logarithmic Paper—Graph paper, also called *ratio paper,* which is so constructed that equal intervals on the vertical scale represent equal rates of change, while equal intervals on the horizontal scale represent equal amounts of change. For instance, the series 1, 2, 4, 8, 16, and 32 would plot as points on a straight line.

Sensitivity Experiment—An experiment in which it is impossible to make more than one observation on a given specimen and, hence, it is impossible to measure a continuous variable of interest. This kind of situation arises, for example, when one wants to determine the maximum thermal shock that can be absorbed by a ceramic product. Once a test has been performed on a given specimen, it is either destroyed or weakened so that it cannot be used again. *See also* Up-and-Down Method

Sequential Sampling—In sequential sampling, observations are made one at a time and a decision is made after each observation whether to accept the hypothesis being tested, whether to reject it, or whether to continue sampling. This kind of sampling often requires substantially fewer observations than would be required (for the same control over Type I and Type II errors) if the sample size were fixed in advance. *See also* Acceptance Number

Serial Correlation—In time series analysis, a serial correlation of order k is the correlation between the values of a time series at times t_i and its values at times t_{i+k}. In the continuous case, it is the correlation between the values of a series at times t and its values at times $t + \tau$, and it is also referred to as an *autocorrelation. See also* Correlogram

Service Rate, Mean—In queuing theory, the average number of units which can be served per unit time; hence, the reciprocal of the average time it takes for a unit to be serviced.

Service Time Distribution – In queuing theory, the distribution of the time it takes for a unit to be served; perhaps, the most common model for service time distributions is the exponential distribution.

Set – A collection, class, or aggregate of objects, real or abstract; the objects which constitute a set are referred to as its *elements*.

Shapley Solution of *n*-Person Game – *See* Solution of a Game

Sheppard's Correction – A correction (adjustment) applied to the calculated values of the moments of a frequency distribution in order to correct for grouping, namely, to correct for the error introduced by the assumption that all the values within a class are located at the midpoint of that class. For details see the book by G. U. Yule and M. G. Kendall listed on page 195.

Sigma, σ – *See* Standard Deviation

Sigma, Σ – Summation sign; $\sum_{i=1}^{n} x_i$ (or simply Σx) represents the sum $x_1 + x_2 + \ldots + x_n$.

σ-Algebra – *See* σ-Field

σ-Field – Also called a *σ-Algebra* or a *Borel Field*, a nonempty class of subsets of a sample space S which is closed under all countable set operations (union, intersection, and complementation). Such fields are used to specify what subsets of a sample space are to be regarded as events in the general definition of probability. For detailed information, see the book by S. S. Wilks listed on page 195.

Sign Tests – Nonparametric tests in which measurements, or differences between measurements, are replaced by plus and minus signs, and then looked upon as a sample from a binomial population. In the *one-sample sign test* of the null hypothesis that the mean of a symmetrical population equals μ_0, each observation exceeding μ_0 is replaced by a plus sign, each observation less than μ_0 is replaced by a minus sign, each observation equaling μ_0 is discarded, and one then tests the null hypothesis that these + and − signs constitute a sample from a binomial population with $\theta = 1/2$. **H.1.** In the *two-sample sign test* (or the *paired samples sign test* when the data are given as matched pairs) of the null hypothesis that two samples come from identical populations, the same technique is applied to the differences between paired data, where the pairing is either random or natural. **H.2** and **H.3**.

Signal Processing Techniques – A collection of mathematical, statistical, and computer techniques used to facilitate the orderly

Signal Processing Techniques – (continued) acquisition, reduction, and analysis of the great masses of data generated in experiments of various sorts (vehicle flight tests, pharmacological research, communications research, and wind-tunnel experiments, for example). Because of their special nature, the computer aspects of signal processing problems are most efficiently dealt with by hybrid computing systems.

Signal-to-Noise Ratio – In communication theory, the ratio of the mean of a random variable to its standard deviation; hence, the reciprocal of the coefficient of variation multiplied by 100.

Signed-Rank Test – *See* Wilcoxon Test

Significance Level – *See* Level of Significance

Significance Test – In hypothesis testing, a test which provides a criterion for deciding whether a difference between theory and practice (a difference between observations and corresponding expectations, or a difference between an observed value of a statistic and an assumed value of a parameter) can reasonably be attributed to chance. If the difference is so small that it can be attributed to chance, one has the option of accepting the hypothesis on which the theoretical value (or values) was based, or of *reserving judgment* (when feasible) by merely stating that the data do not permit the rejection of the null hypothesis.

Simple Aggregative Index – An index constructed for an aggregate, or collection of items, which is given by the *ratio* of the sum of their given-year prices (quantities, or values) to the sum of their base-year prices (quantities, or values); this ratio is usually multiplied by 100 to express the index as a percentage. **K.3.**

Simple Averages, Method of – In time series analysis, a very simple, and also very unsophisticated method of describing a seasonal pattern; that is, constructing a seasonal index. Nowadays, the method is rarely used because it assumes an additive rather than a multiplicative model.

Simple Hypothesis – A hypothesis is said to be simple if it completely specifies the distribution of a random variable. For example, the hypothesis that the mean and the standard deviation of a normal population are, respectively, $\mu = \mu_0$ and $\sigma = \sigma_0$ is a simple hypothesis. *See also* Composite Hypothesis

Simple Random Sample – *See* Random Sample

Simplex Method – A systematic iterative computational procedure devised originally (by G. B. Dantzig) to solve linear programming problems. Beyond its usefulness in these problems and in solving finite zero-sum two-person games, the simplex method is the best general method available for finding non-negative solutions of

Simplex Method—(continued)
systems of linear equations and linear inequalities. A simple illustration of this method may be found in the book by M. Sasieni, A. Yaspan, and L. Friedman listed on page 195.

SIMSCRIPT—A computer programming language designed especially to facilitate the writing of simulation programs. Besides its applicability in building almost any kind of simulation model, SIMSCRIPT is a general programming system which may be advantageously used in problems of optimization, statistical analysis, data reduction, and so on.

Simulation—The artificial generation of random processes (usually by means of random numbers and/or computers) to imitate or duplicate actual physical processes. *See also* Monte Carlo Methods

Single Sampling—*See* Acceptance Number

Single-Tail Test—*See* One-Tail Test (One-Sided Test)

Size of Critical Region—The probability α of committing a Type I error. *See also* Critical Region

Size of a Population—The number of elements in a finite population; it is usually denoted by the letter N.

Size of a Sample—The number of observations in a sample; it is usually denoted by the letter n.

Skewness—The lack of symmetry in a distribution; it is usually measured by the statistic α_3 (*alpha-three*), whose value is zero for a symmetrical distribution, or by the *Pearsonian Coefficient of Skewness*, whose value is zero when the median of a distribution coincides with its mean. **A.18, A.20, and E.13.** A distribution is said to have *positive skewness* (or be *positively skewed*) when it has a long thin tail at the right, and α_3 as well as the Pearsonian coefficient of skewness are, in fact, positive. A distribution is said to have *negative skewness* (or be *negatively skewed*) when it has a long thin tail at the left, and α_3 as well as the Pearsonian coefficient of skewness are, in fact, negative.

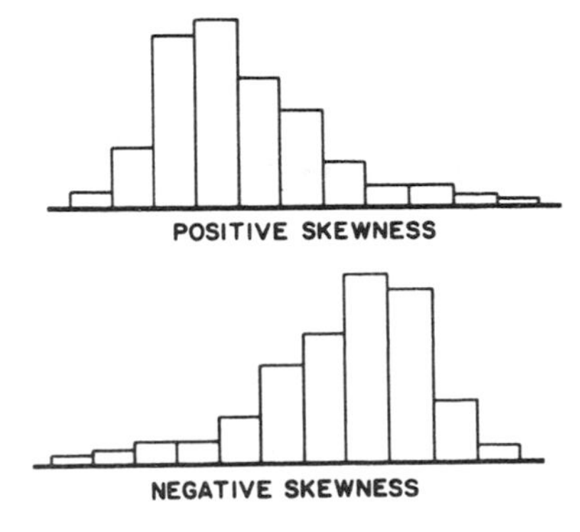

Smoothing—In time series analysis, the removal of minor fluctuations or erratic fluctuations from a series of data. This is often accomplished by use of a moving average.

Software—All of the programming aids (such as programming languages and various operating routines), usually supplied by computer manufacturers, intended to help users make the most efficient use of their computer hardware. For example, the FORTRAN system is a part of the software package supplied by many manufacturers. *See also* Hardware (of a Computer)

Solution of a Game—The solution of a finite zero-sum two-person game consists of a set of optimum strategies for each player and the value of the game, that is, the payoff or expected payoff corresponding to these optimum strategies. For n-person games, there are various concepts concerning the solution of a game; for example, the *von Neumann concept* and the *Shapley concept.* For details see the book by E. Burger listed on page 193. *See also* Imputation

Source Language—The language, such as FORTRAN or COBOL, in which a source program is written.

Source Program—A computer program written in a language other than machine language. Source programs are ordinarily translated into object programs by some sort of automatic coding system.

Spearman-Brown Formula—In psychology and education, a formula which enables one to estimate the reliability of a long test from the known reliability of a shorter test to which the longer test is comparable in every other way. This formula may be found in the second book by A. L. Edwards listed on page 193.

Spearman's Rank Correlation Coefficient—A measure of rank correlation which, when there are no ties, equals the coefficient of correlation calculated for the ranked data. **I.29.** *See also* Kendall's Tau; Rank Correlation

Special Purpose Index—*See* General Purpose Index

Special Rule of Addition—*See* General Rule of Addition

Special Rule of Multiplication—*See* General Rule of Multiplication

Specific Seasonal Pattern—In time series analysis, a pattern, resulting from the impact of seasonal forces, which is observed in a series of data for one period; strictly speaking, the twelve monthly seasonal values observed for a single year. *See also* Typical Seasonal Pattern

Split-Half Method—In psychology and education, a procedure for determining the reliability of various kinds of tests by correlating "split halves;" that is, the scores on the odd items and on the even items, or the scores on the first half and the second half of a test.

Split-Plot Design—An experimental design in which a main effect (or a set of treatments) is confounded with blocks, in the sense that the

Split-Plot Design – (continued) experimental units for this effect (or these treatments) are subdivided into subplots for the allocation of another main effect (or a second set of treatments.)

Spot Market Prices, Index of – A daily index (also published in the form of monthly averages) designed to measure the price movement of twenty-two sensitive basic commodities (nine foodstuffs and thirteen raw materials) whose markets are, presumably, among the first to be affected by actual or anticipated changes in economic conditions. The index is prepared by the U.S. Department of Labor.

Spurious Correlation – (1) This term is sometimes used when a high positive or negative value obtained for the coefficient of correlation can be attributed entirely to chance. (2) The term is also used to describe the inflated correlation which results from an overlap of two variables that are being correlated (for example, when scores on two tests consisting partly of the same items are correlated).

Square Root Transformation – The transformation $y = \sqrt{x}$ or $y = \sqrt{x + 1/2}$ is often used to make data consisting of values of random variables having Poisson distributions amenable to analysis of variance techniques.

(s,S) Policy – An inventory ordering policy which specifies two numbers s and S, with $0 \leq s \leq S$, and requires that when the stock level x of an item declines below s, a replenishment order for $S - x$ items be placed so as to bring the stock level up to S. The predetermined quantity s is called the *reorder point* or *trigger level.*

Standard Deviation – (1) The standard deviation of a sample of size n (usually called a "sample standard deviation") is given by the square root of the sum of the squared deviations from the mean divided by $n - 1$. **A.10** and **B.8**. It is by far the most widely used measure of the variation of a set of data and it is generally denoted by the letter s. Some statisticians prefer to divide by n (rather than by $n - 1$) in the definition of the standard deviation, in which case it has also been referred to as the *root-mean-square deviation* and it may be described as the square root of the second moment about the mean. The square of the sample standard deviation is called the *sample variance.* (2) The standard deviation of a finite population is defined in the same way as the standard deviation of a sample; it is generally denoted by the Greek letter σ (*sigma*) when division is by the population size N, and by S when division is by $N - 1$. **C.2.** The square of the population standard deviation is called the *population variance.* (3) The standard deviation of the distribution of a random variable is given by the square root of the second moment about the mean and it is generally denoted by the Greek letter σ; such a standard deviation

Standard Deviation – (continued)
is usually referred to as a *population standard deviation,* and its square as a *population variance.* E.6 and E.7.

Standard Error – The standard deviation of the sampling distribution of a statistic. I.23 for the coefficient of correlation, F.4 for the difference between two means, G.3 for the difference between two proportions, F.1 - F.3 for the mean, F.5 for the median, G.1 and G.2 for a proportion, F.6 for the standard deviation, and F.7 for the variance.

Standard Error of Estimate – In regression analysis, the square root of the residual variance.

Standard Measure – *See* Standard Score

Standard Normal Distribution – *See* Normal Distribution

Standard Score – A standard score, *z-score,* or *standard measure* is a score converted into standard units. *See also* Normalized Standard Scores; *T*-Scores

Standard Units – A set of data is converted into standard units by subtracting from each value the mean of the data and then dividing by their standard deviation. Note that in standard units the mean of any set of data is zero and its standard deviation is equal to 1.

Standardized Mean – The random variable $Y = \sqrt{n}\,(\bar{X} - \mu)/\sigma$, where μ and σ are, respectively, the mean and the standard deviation of the population from which the sample was obtained.

Standardized Random Variable – A random variable which has been transformed linearly so that its mean is 0 and its standard deviation is 1. If X and Y are a random variable and the corresponding standardized random variable, then $Y = (X - \mu)/\sigma$, where μ and σ are, respectively, the mean and the standard deviation of the distribution of X.

Standardized Test – A test whose items have been carefully selected and evaluated, and which is accompanied by such things as directions for its administration and scoring, tables of norms, and suggestions for interpreting the results.

Stanine Scale – A nine-point scale of normalized standard scores whose values 1, 2, . . . , 9 have a mean of 5 and a standard deviation of 2. The word "stanine," applied to an individual value in the scale, is a contraction of "standard nine."

Stationary Process – A random process whose statistical characteristics remain unchanged as time progresses; a more formal definition may be found in the book by E. Parzen listed on page 195. A random process which is not stationary is referred to as *evolutionary.* For example, if a random process consists of a sequence of 0s and 1s representing "failures" and "successes" in a sequence of independent Bernoulli

Stationary Process – (continued)
trials with equal probabilities of a "success," this process is stationary; however, if one considers the accumulated number of successes, this process is evolutionary since the distribution of the total number of successes depends on the number of trials.

Statistic – A quantity (such as a mean or a standard deviation) calculated on the basis of a sample. In practice, the term is used for values of such random variables as well as the random variables themselves.

Statistical Control – In quality control, a process is said to be in statistical control when all sources of assignable variation have been removed; that is, when the variability of the process is confined to chance variation.

Statistical Hypothesis – *See* Hypothesis

Statistical Inference – Also called inductive statistics, a form of reasoning from sample data to population parameters; that is, any generalization, prediction, estimate, or decision based on a sample. It is customary to refer to the Neyman-Pearson theory of testing hypotheses and the method of confidence intervals as classical statistical inference. *See also* Bayesian Inference; Fiducial Probability

Statistics – (1) The totality of methods employed in the collection and analysis of any kind of data and, more broadly, that branch of mathematics which deals with all aspects of the science of decision making in the face of uncertainty. (2) A collection of numerical data such as those found in the financial pages of newspapers, the *Statistical Abstract of the United States*, Census Reports, and the like. The word "statistics" comes from the Latin word *status*, meaning a political state.

Steepest Ascent, Method of – In response surface analysis, a sequential method of experimentation for seeking optimum conditions.

Stochastic Convergence – A sequence of random variables (for example, a sequence of sample proportions calculated for a sequence of successive trials) is said to converge stochastically, or *in probability*, if it converges in the sense of the weak law of large numbers.

Stochastic Independence – This term is synonymous with "independence," as defined in probability theory. *See also* Independent Events; Independent Random Variables

Stochastic Matrix – *See* Markov Chain

Stochastic Process – *See* Random Process

Stochastic Variable – *See* Random Variable

Storage Section (of a Digital Computer) – *See* Digital Computer

Strata – *See* Stratified Random Sampling

Strategy—In game theory, a strategy is a plan which specifies completely what choices a player should make in all circumstances that can possibly arise during the course of play. If a strategy calls for a specific move (or a specific sequence of moves) in each situation that might arise, the strategy is referred to as *pure;* if a strategy allows for a choice among several moves (or courses of action) and this choice is somehow controlled by chance, the strategy is referred to as *randomized* or *mixed.*

Stratified Random Sampling—A method of sampling in which portions of the total sample are allocated to individual subpopulations and randomly selected from these *strata.* The principal purpose of this kind of sampling is to guarantee that population subdivisions of interest are represented in the sample, and to improve the precision of whatever estimates are to be made from the sample data. F.3. *See also* Equal Allocation; Optimum Allocation; Proportional Allocation

Strictly Determined Game—*See* Saddle Point

Student-*t* Distribution—*See* *t* Distribution

Studentization—The removal of a nuisance parameter by constructing a statistic whose sampling distribution does not depend on this parameter. This idea was first used by W. S. Gossett (who used the pen name "Student") in connection with the sampling distribution of the mean.

Studentized Mean—The statistic $t = \sqrt{n}\,(\bar{x} - \mu)/s$, whose sampling distribution for random samples from normal populations (unlike that of $\bar{x}$ alone) does not depend on the population standard deviation.

Sturges' Rule—A rule for determining the number of classes into which a set of n numbers might most profitably be grouped. The formula for the number of classes is $1 + 3.3(\log n)$; it is nowadays rarely used.

Subjective Probability—A point of view which is currently gaining in favor is to interpret probabilities (particularly those relating to single, nonrepetitive events) as subjective, or *personal,* probabilities; namely, as measures of the strength of a person's belief concerning the occurrence or nonoccurrence of events. The use of subjective probabilities is advocated, in particular, in conjunction with methods of Bayesian inference.

Subroutine—A portion of a complete computer routine, which consists of a set of instructions which cause the machine to perform some well-defined mathematical or logical operation; for example, compute a square root or find a sum of squares.

Subsampling—The process of selecting a sample from a sample. *See also* Master Sample; Multi-Phase Sampling; Multi-Stage Sampling

Subset—Set A is a subset of set B if and only if each element of A is also an element of B. Thus, each set is a subset of itself and, by

Subset–(continued)
definition, the empty set $\emptyset$ which has no elements at all is a subset of every set. *See also* Event

Sufficiency–An estimator is said to be sufficient if it utilizes all the information in a sample that is relevant to the estimation of a given parameter. More rigorously, an estimator of a given parameter is sufficient if the conditional likelihood of the sample *given the value of the estimator* does not depend on the parameter.

Sum of Squares–In analysis of variance, a sum of the squared deviations of observed quantities from appropriate means. *See also* Between-Samples Sum of Squares; Block Sum of Squares; Error Sum of Squares; Total Sum of Squares; Treatment Sum of Squares; Within-Samples Sum of Squares

Summation Sign–*See* Sigma, Σ

Survey–*See* Sample Design; Sample Survey

Survey Design–*See* Sample Design

Symmetrical Distribution–A distribution is symmetrical if values equidistant from the mean have identical frequencies, probabilities, or probability densities. The distribution shown in the figure on page 9, for example, is symmetrical. *See also* Skewness

System, Computer Operating–A comprehensive collection of control programs and language processing programs (such as FORTRAN, COBOL, and PL/1 compilers) which enables a digital computer to direct and control its own operations. By automatically selecting and assigning input-output devices, selecting and loading programs from the program library, and compiling and executing applications programs, and so on, powerful operating systems such as those for the IBM System/360 greatly increase the efficiency of a computer facility.

System, Control–In the theory of control processes, a system is any well-defined process or activity which relates a quantity to be controlled to a quantity which may be observed. The controlled quantity is called the *input* and the observed quantity the *output* (or *response*) of the system. The control problem is essentially that of finding the relationship between input, output, and the system. *See also* Cybernetics

System of Equations–A set of equations to be solved for values of the variables which satisfy all of the equations. *See also* Determinant; Linear Equation; Normal Equations

Systematic Error–A nonrandom error which introduces a bias into all the observations; such an error might be caused by faulty, or poorly adjusted, measuring instruments.

Systematic Sample—A sample obtained by selecting every kth item on a list, every kth voucher in a file, every kth house on a street, every kth piece coming off an assembly line, and so on. An element of randomness may be introduced into this kind of sampling by randomly selecting from the first k units the unit with which to start. This is referred to as a *random start*, and a sample so chosen is sometimes called an "every kth" systematic sample.

T

t **Distribution**—A distribution which is used largely for inferences concerning the mean (or means) of normal distributions whose variances are unknown. E.37. It is also referred to as the *Student-t distribution*, after W. S. Gossett, who used the pen name "Student." The parameter of this distribution is ν, the number of degrees of freedom. The table on page 109 contains values of $t_{\alpha,\nu}$, which denotes the value for which the area *to its right* under the t distribution with ν degrees of freedom is equal to α.

t **Tests**—Tests based on statistics having the t distribution. The *one-sample t test* is a test of the null hypothesis that a random sample comes from a normal population with the mean $\mu = \mu_0$; it is used when the population standard deviation is unknown. F.18. The *two-sample t test* is a test concerning the difference between the means of two normal populations having the same standard deviation; it is based on independent random samples from the two populations. F.20. The *paired-samples t test* is an application of the one-sample t test to differences between paired data; it is used when the data are actually given as *matched pairs* or when they are *randomly matched* because the assumption of equal standard deviations in the two-sample t test cannot be met.

T **Scores**—In psychology and educations scores that are used in constructing norms for standardized tests; they are normalized standard scores which are linearly transformed so that their mean is 50 and their standard deviation is 10. Thus, the formula for this transformation is $T = 50 + 10$ (normalized standard score). *See also* Normalized Standard Scores

Test—(1) A decision procedure for accepting or rejecting a statistical hypothesis. (2) An examination. *See also* Hypothesis; Neyman-Pearson Theory; Significance Test

Test Battery—A collection of tests (examinations), each of which measures some aspect of the traits in question. For example, there are various test batteries for measuring attainment in different elementary school subjects.

THE t DISTRIBUTION*

The entries in this table are values of $t_{\alpha,\nu}$, for which the area to their right under the t distribution with ν degrees of freedom is equal to α. For $\alpha > 0.50$ it is necessary to make use of the identity $t_{\alpha,\nu} = {}^{-}t_{1-\alpha,\nu}$.

ν	$\alpha = .10$	$\alpha = .05$	$\alpha = .025$	$\alpha = .01$	$\alpha = .005$	ν
1	3.078	6.314	12.706	31.821	63.657	1
2	1.886	2.920	4.303	6.965	9.925	2
3	1.638	2.353	3.182	4.541	5.841	3
4	1.533	2.132	2.776	3.747	4.604	4
5	1.476	2.015	2.571	3.365	4.032	5
6	1.440	1.943	2.447	3.143	3.707	6
7	1.415	1.895	2.365	2.998	3.499	7
8	1.397	1.860	2.306	2.896	3.355	8
9	1.383	1.833	2.262	2.821	3.250	9
10	1.372	1.812	2.228	2.764	3.169	10
11	1.363	1.796	2.201	2.718	3.106	11
12	1.356	1.782	2.179	2.681	3.055	12
13	1.350	1.771	2.160	2.650	3.012	13
14	1.345	1.761	2.145	2.624	2.977	14
15	1.341	1.753	2.131	2.602	2.947	15
16	1.337	1.746	2.120	2.583	2.921	16
17	1.333	1.740	2.110	2.567	2.898	17
18	1.330	1.734	2.101	2.552	2.878	18
19	1.328	1.729	2.093	2.539	2.861	19
20	1.325	1.725	2.086	2.528	2.845	20
21	1.323	1.721	2.080	2.518	2.831	21
22	1.321	1.717	2.074	2.508	2.819	22
23	1.319	1.714	2.069	2.500	2.807	23
24	1.318	1.711	2.064	2.492	2.797	24
25	1.316	1.708	2.060	2.485	2.787	25
26	1.315	1.706	2.056	2.479	2.779	26
27	1.314	1.703	2.052	2.473	2.771	27
28	1.313	1.701	2.048	2.467	2.763	28
29	1.311	1.699	2.045	2.462	2.756	29
∞	1.282	1.645	1.960	2.326	2.576	∞

*Abridged from Table IV of R. A. Fisher, *Statistical Methods for Research Workers*, published by Oliver and Boyd, Ltd., Edinburgh, by permission of the author's literary executor and publishers.

Test of Significance – *See* Significance Test

Test Statistic – A statistic on which the decision whether to accept or reject a given hypothesis is based.

Tests of Hypotheses – Rules, or procedures, for deciding whether to accept or reject a hypothesis. *See also* Hypothesis; Neyman-Pearson Theory; Significance Test

Tetrachoric Correlation – A measure of association in a *double dichotomy* (2 by 2 table), which applies when both of the dichotomized variables are actually continuous and normally distributed. **G.18.**

Theoretical Distribution – A general term used to denote the distribution of a random variable, as contrasted with a distribution of observed data.

Theoretical Frequency – In some applications, an alternate term for "expected frequency."

Theory of Games – *See* Game Theory; Game of Strategy

Three-Sigma Control Limits – *See* Control Chart

Ties (in Rank) – In most statistical applications, ties in rank are resolved by assigning each of the items which are tied the *mean* of the ranks they jointly occupy. For example, if two items are tied for third and fourth, they are both assigned a rank of $3\frac{1}{2}$; if three items are tied for fourth, fifth, and sixth, they are all assigned a rank of 5.

Tightened Inspection – *See* Normal Inspection

Time Reversal Test – A test of index number quality which requires that if the functions of the given year (or period) and the base year (or period) are interchanged (that is, if the subscripts o and n are interchanged in an index number formula wherever they appear), the new index be the reciprocal of the original index (expressed as a proportion). The Laspeyres' index, for example, does not meet the time reversal test, but Fisher's Ideal Index does.

Time Series – Any series of data collected, observed, or recorded at regular intervals of time; for instance, the grape tonnage harvested in Napa County, California, during the month of October for the years 1946 through 1965. Though usually applied to economic and business data, the term and the techniques developed for analyzing such data apply also to the treatment of data from any of the social and natural sciences.

Time Series Analysis – The study of time series, or more specifically, the separation or decomposition of a time series into its individual components according to some model (or some set of assumptions). The ultimate goal of most analyses of time series is to make predictions or forecasts. *See also* Multiplicative Model

Tolerance Limits – In industrial applications, values between which one can expect to find a given proportion of a population. Whenever such values are determined on the basis of samples, one can only assert with a specified probability (or degree of confidence) that *at least* a given proportion of the population falls between the limits.

Total Information – In quality control, a term used to designate all of the information contained in an observed set of numbers which have been arranged in ascending order of size. *See also* Essential Information

Total Inspection – An alternate term for 100 per cent inspection.

Total Sum of Squares – In analysis of variance, the sum of the squared deviations of all the observations (in a given experiment) from their mean. The foremost objective of an analysis of variance is to divide the total sum of squares into components which can be attributed to various specific sources of variation. (J.1c), (J.2b), and (J.3c).

Transformation – A change of variable. Sometimes, transformations are performed to simplify calculations; sometimes, random variables are transformed so that one can meet the assumptions underlying "standard" methods; and sometimes random variables are transformed so that their distributions are of a certain well-known type. *See also* Arc Sine Transformation; Coding; Square Root Transformation; z-Transformation

Transition Matrix – *See* Markov Chain

Transition Probability – *See* Markov Chain

Treatment – A treatment is an experimental condition; for example, different treatments can be different levels of a factor (or different values of a variable), or they can be different combinations of the levels of several factors. In other words, the term "treatment" may denote, literally, two fertilizer treatments for corn or four diets for animals, but also three machines operated at the same speed or the nine combinations possible when a machine is operated by each of three persons at each of three speeds.

Treatment Effect – In analysis of variance, a quantity (usually a parameter) which represents the change in response produced by a given treatment. In (J.1a) and (J.3a) the treatment effects are the parameters α_i.

Treatment Sum of Squares – In analysis of variance, that component of the total sum of squares which can be attributed to possible differences among the treatments. (J.1d), (J.2c), and (J.3d).

Tree Diagram – A diagram in which the individual paths (all emanating from one point) represent all possible outcomes of an experiment. Such diagrams are of special value in situations where the end-result is attained through various intermediate steps.

Trend – *See* Secular Trend

Trend-Cycle Component – When treated as a single component in time series analysis, it is considered to be the combination of the underlying long-term trend, the periodic movements that accompany economic cycles, and the short-term subcycles that have occurred in a series.

Trend Increment – *See* Annual Trend Increment; Monthly Trend Increment

Trial – This term is used to designate one of a series of repeated experiments, such as repeated flips of a coin, where one is interested, for instance, in the probability of getting x "successes" in n "trials." *See also* Bernoulli Trial

True Mean – The mean of a population; the term is meant to emphasize the distinction between a sample mean and the constant (though unknown) mean of a population.

True Regression – A regression (equation) expressed in terms of the parameters of the model. For example, **I.8**. The term is meant to emphasize the distinction between such a regression equation and an *estimated* regression equation, where the regression coefficients are replaced by estimates.

Truncated Data – This term is used when sample values exceeding a fixed constant and/or less than a fixed constant are *not* observed or recorded; the constant (or constants) are referred to as *points of truncation* and the proportion of values thus excluded is sometimes called the *degree of truncation*. *See also* Censored Data

Truncated Distribution – A distribution is said to be truncated if it is modified so that values exceeding a fixed constant and/or less than a fixed constant are excluded. For example, the binomial and Poisson distributions are sometimes truncated by excluding the possibility of zero successes; one thus obtains the probabilities of x successes *given that there is at least one.*

Tukey's Test – *See* Multiple Comparisons Tests

Two-Person Game – *See* Game of Strategy

Two-Phase Sampling – *See* Multi-Phase Sampling

Two-Sample Sign Test – *See* Sign Tests

Two-Sample t Test – *See* t Tests

Two-Sided Alternative – *See* One-Sided Alternative

Two-Stage Sampling – *See* Multi-Stage Sampling

Two-Tail Test – *See* One-Tail Test (One-Sided Test)

Two-Way Analysis of Variance – An analysis of variance, where the total sum of squares is expressed as the sum of the treatment sum of squares, the block sum of squares, the error sum of squares, and no others. **J.3.**

Two-Way Classification – A classification of a set of observations according to two characteristics (or two variables). Thus, the term is used in analysis of variance when the data are grouped according to treatments as well as blocks. *See also* Correlation Table; Dichotomy

Type I Error – In hypothesis testing, the erroneous rejection of a null hypothesis; the probability of committing a Type I error is usually denoted by α. *See also* Producer's Risk

Type II Error – In hypothesis testing, the erroneous acceptance of a null hypothesis; the probability of committing a Type II error is usually denoted by β. *See also* Consumer's Risk; Power; Power Function

Typical Seasonal Pattern – A pattern obtained by averaging in some way the specific seasonal patterns observed (or calculated) for a number of years. The twelve numbers which result from averaging the specific Januaries together, and so on through the specific Decembers, are called a "seasonal index." *See also* Seasonal Index; Specific Seasonal Pattern

U

U-Shaped Distribution – A frequency distribution having the general shape of the letter U.

U-Shaped Distribution

U-Test – *See* Mann-Whitney Test

Unbiased Estimator – An estimator whose expected value (namely, the mean of its sampling distribution) equals the parameter it is intended to estimate. *See also* Bias

Uniform Distribution – A distribution which in the *discrete case* assigns the same probability to each value within its domain; in the *continuous case* it has a constant probability density over a given interval. E.24 and E.38. Uniform distributions are also referred to as *rectangular distributions.*

Uniformly Better Decision Function – One decision function is uniformly better than another according to some criterion if it is sometimes better but never worse than the other according to this criterion.

Uniformly More Powerful Test—One test of a given hypothesis is uniformly more powerful than another if it is at least as powerful as the other for all values of the parameter under consideration, and more powerful for at least one value of the parameter. *See also* Power; Power Function

Uniformly Most Powerful Test—A test of a given hypothesis is uniformly most powerful if it is uniformly more powerful than any other test having a critical region of the same size α. *See also* Power; Power Function; Uniformly More Powerful Test

Unimodal—A set of data or a bistribution is said to be unimodal if it has only one mode or modal class. *See also* Multi-Modal Distribution

Union (of Two Sets)—The union of two sets A and B, denoted $A \cup B$, is the set which consists of all elements that belong to A, to B, or to both.

Unit of Information—*See* Bit

Units Test—A test of index number quality which requires that a price index be independent of the units (ounces, pounds, tons, etc.) to which the prices refer. This criterion is met by most weighted indexes, but not by a simple aggregative index.

Univariate Distribution—The distribution of one random variable.

Universal Turing Machine—Any computer which has the capacity of imitating the behavior of any other such machine, regardless of the symbols read or written or the number of internal states in the machine being simulated. Described in 1936 by A. M. Turing, such machines may be used to simulate other digital systems, or for computations. Present-day digital computers (provided they have sufficient storage capacity) are Universal Turing Machines.

Universe—A synonym for "population;" it is nowadays rarely used.

Unweighted Index Number—An index number in the calculation of which no overt weights are used in averaging the prices or price relatives (quantities or quantity relatives).

Up-and-Down, Runs—*See* Runs Up-and-Down

Up-and-Down Method—A technique of conducting sensitivity experiments; depending on the outcome of each test, the level of the variable under consideration is lowered or raised for the next test. A discussion of this method may be found in the book by W. J. Dixon and F. J. Massey listed on page 193. *See also* Sensitivity Experiment

Upper Control Limit, UCL—*See* Control Chart

Upward Bias—The inherent tendency of an estimator or statistical measure to overestimate or overstate the phenomenon it is intended to describe.

Utile—A unit of utility.

Utility, Expected—*See* Utility Function

Utility Function—A function which, for a given individual, associates real numbers with consequences (outcomes, or prospects) such that (1) the utility of consequence C_1 is greater than that of consequence C_2 if, and only if, the individual prefers C_1 to C_2, and (2) if C is the consequence where the individual faces C_1 and C_2 with respective probabilities of p and $1 - p$, then the utility of C equals p times the utility of C_1 plus $1 - p$ times the utility of C_2. If a utility function is defined over a sample space, the expected value of this random variable is referred to as the corresponding *expected utility*. Expected utilities are often calculated by means of formulas analogous to **D.13** (the formula for a *mathematical expectation*), letting a_i be the utility an individual associates with outcome A_i.

V

Validity—In psychology and education, the extent to which a test measures what it purports to measure. *See also* Validity Coefficient

Validity Coefficient—In the construction of psychological tests, the correlation between scores on a test and some independent measure of the quantity the test purports to measure.

Value of a Game—In a zero-sum two-person game, the payoff (or expected payoff) which corresponds to optimum strategies for both players. *See also* Solution of a Game

Value Index—An index number reflecting changes in both physical volume and prices; for example, changes in total sales, inventories, or wages.

Variables, Inspection by —In quality control, inspection in which quality characteristics of an item are determined by measuring them on a continuous scale; for example, by measuring the weight of a package in ounces. *See also* Attributes, Inspection by

Variance—*See* Standard Deviation

Variance-Components Model—*See* Random-Effects Model

Variance-Covariance Matrix—In multivariate analysis, a matrix for which the element a_{ij} is given by the covariance of the ith and jth random variables when $i \neq j$, and by the variance of the ith random variable when $i = j$.

Variance Ratio—An alternate name for a statistic having the F distribution; specifically, the ratio of two independent estimates of a population variance.

Variance-Ratio Distribution—*See* F Distribution

Variate—*See* Random Variable

Variation— The extent to which observations or distributions are spread out, or dispersed. *See also* Chance Variation

Variation, Coefficient of — A widely used measure of *relative variation* which is given by the ratio of the standard deviation of a set of data to their mean, thus expressing the magnitude of their variation relative to their average size. This ratio is often multiplied by 100 to express the measure of relative variation as a percentage. A.15. *See also* Quartile Variation, Coefficient of; Signal-to-Noise Ratio

Variation, Measures of— Statistical descriptions such as the standard deviation, the mean deviation, or the range, which are indicative of the spread, or dispersion, of a set of data or distribution.

Vector— A matrix having only one row or one column, for example, $\mathbf{x} = (x_1, x_2, \ldots, x_n)$; its elements are referred to as the *components* of the vector, and the integer n is called the *dimension* of the vector. An ordered n-tuple of random variables, looked upon as the components of a vector, is sometimes referred to as a *random vector*.

Venn Diagram— A diagram in which sets are represented by circular regions, parts of circular regions, or their complements with respect to a rectangle representing the sample space which, in this connection, is also referred to as a "universal set" or a "universe of discourse." Venn diagrams are often used to verify relationships among sets, subsets (or events).

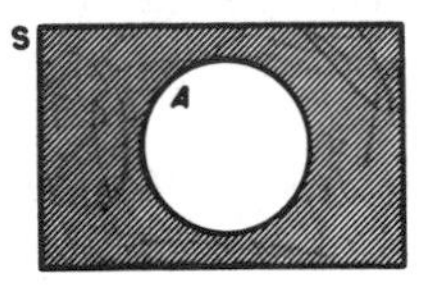

A' Shaded

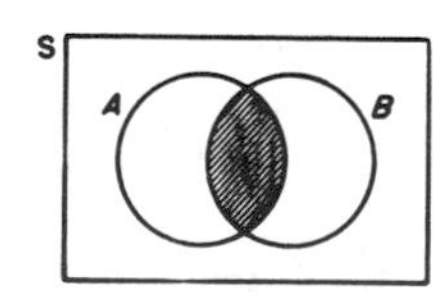

$A \cap B$ Shaded

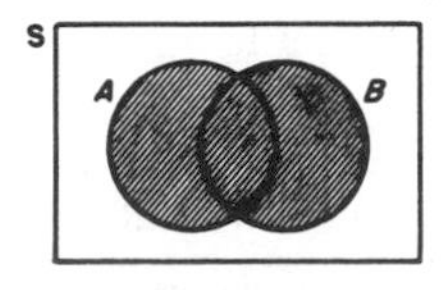

$A \cup B$ Shaded

Venn Diagrams

Von Neumann Solution of *n*-Person Game – *See* Solution of a Game

W

Waiting Line – *See* Queue

Waiting Line Theory – *See* Queuing Theory

Wald-Wolfowitz Test – A nonparametric test of the null hypothesis that two samples come from identical populations; it is based on the theory of runs. **H.10.**

Weibull Distribution – A distribution which is used extensively in the study of the reliability of industrial products. **E.39.** An important feature of this distribution is that the corresponding failure rate can be made to increase with time, decrease with time, or remain constant, by a proper choice of the parameters. In the special case where the failure rate is constant, the Weibull distribution reduces to the exponential distribution.

Weighted Aggregative Index – An index number constructed for an aggregate, or collection, of items which have been weighted in some way so as to reflect their relative importance with regard to the over-all phenomenon the index is designed to describe. In a price index the weights are usually the corresponding quantities produced, consumed, sold, etc. **K.7.** *See also* Fixed-Weight Aggregative Index; Laspeyres' Index; Paasche's Index

Weighted Index Number – *See* Fixed-Weight Aggregative Index; Laspeyres' Index; Paasche's Index; Weighted Aggregative Index; Weighted Mean of Price Relatives

Weighted Mean – The average of a set of numbers obtained by multiplying each number by a weight expressing its relative importance, and then dividing the sum of these products by the sum of the weights. **A.2.** For instance, if one buys five boxes of berries at 49 cents a box and two more boxes at 70 cents a box, the (weighted) mean price per box is $(49 \cdot 5 + 70 \cdot 2)/7 = 55$ cents.

Weighted Mean of Price Relatives – A price index given by a weighted mean of price relatives; the weights are usually the total values of the corresponding items produced, consumed, sold, etc., in the base year (or period) or in the given year (or period). **K.6.**

Weighted Moving Average – A moving average in which the individual terms are explicitly weighted so as best to accomplish the purpose for which the moving average is calculated (usually to eliminate irregular variations). Any number of weighting systems are in use (or have been tried) in connection with the smoothing of economic time series, including weights based on binomial coefficients,

Weighted Moving Average—(continued)
parabolic weights, exponential weights, and many others. For further information see the book by R. G. Brown listed on page 193.

Wherry-Doolittle (Test Selection) Method—A modification of the Doolittle method of solving systems of linear equations, the Wherry-Doolittle method selects, from a number of tests tried experimentally, the best battery of tests for predicting a particular criterion (vocational success, for example). The method adds tests in order of their predictive importance, making allowance for the chance error added by each test, until the multiple correlation between the predictors and the criterion is maximized. This method is described in the book by W. H. Stead, C. L. Shartle, et al., listed on page 195.

Wholesale Price Index—Constructed by the Bureau of Labor Statistics, the Wholesale Price Index is intended to measure changes in the prices of large lots of commodities in primary markets; that is, at their first important commercial transaction. The primary source of this important index is the *Monthly Labor Review.*

Wiener-Khintchine Relations—Fundamental relations in the harmonic analysis of stochastic processes. Formal statements of these relations may be found in the book by E. Parzen listed on page 195.

Wiener Process—A random process designed originally as a model for Brownian motion. Nowadays, the process has found many other applications, and details may be found in the book by E. Parzen listed on page 195.

Wilcoxon Test—Also called the *matched-pairs signed-rank test,* the Wilcoxon test is a nonparametric alternative for the paired-sample (or two-sample) t test. **H.7**. The absolute values of the differences between paired observations are ranked, and the test is based on the sum of the ranks of the differences which were originally positive (or negative, whichever sum is smaller). *See also* Mann-Whitney Test

Within-Samples Sum of Squares—Another name for the error sum of squares in a one-way analysis of variance. (**J.1e**).

Y

Youden Square—An incomplete block design which may be obtained by deleting one or more rows from a Latin square and treating the columns as blocks; such a design is also referred to as an *incomplete Latin square.* Information about this kind of design may be found in the book by W. G. Cochran and G. M. Cox listed on page 193.

Yule Process—*See* Pure Birth Process

Z

***z*-Scores** – An alternate name for standard scores, that is, scores converted into standard units.

***z*-Transformation** – A transformation of the sample correlation coefficient by means of the formula $z = \frac{1}{2} \ln \frac{1+r}{1-r}$. **I.22.** This transformation is used to perform significance tests and construct confidence limits for correlation coefficients. **I.24** and **I.26**.

Zero-Sum Game – *See* Game of Strategy

Part II
STATISTICAL FORMULAS

Explanations of the notation used, assumptions underlying the various techniques, and other comments are given at the beginning of each section.

A. THE DESCRIPTION OF UNGROUPED DATA

B. THE DESCRIPTION OF GROUPED DATA

C. THE DESCRIPTION OF FINITE POPULATIONS

D. PROBABILITY

E. PROBABILITY DISTRIBUTIONS

F. FORMULAS FOR INFERENCES CONCERNING MEANS AND STANDARD DEVIATIONS

G. FORMULAS FOR THE ANALYSIS OF COUNT DATA

H. NONPARAMETRIC TESTS

I. CURVE FITTING, REGRESSION, AND CORRELATION

J. ANALYSIS OF VARIANCE

K. INDEX NUMBER FORMULAS

L. FORMULAS FOR CONTROL CHARTS

Part II

STATISTICAL FORMULAS

A. THE DESCRIPTION OF UNGROUPED DATA

The formulas of this section apply to a set of n measurements or observations $x_1, x_2, \ldots,$ and x_n, which are sometimes referred to simply as x's. The appearance of summations is simplified by omitting subscripts and limits; thus,

$$\sum_{i=1}^{n} x_i = x_1 + x_2 + \ldots + x_n \text{ is written } \Sigma x.$$

A.1 **Mean (Arithmetic Mean)**

$$\bar{x} = \frac{x_1 + x_2 + \ldots + x_n}{n} = \frac{\Sigma x}{n}$$

A.2 **Weighted Mean**—If the x's are assigned corresponding weights $w_1, w_2, \ldots,$ and w_n, the weighted mean is given by

$$\bar{x}_w = \frac{x_1 w_1 + x_2 w_2 + \ldots + x_n w_n}{w_1 + w_2 + \ldots + w_n} = \frac{\Sigma x \cdot w}{\Sigma w}$$

A.3 **Geometric Mean**

(A.3a) $$G = \sqrt[n]{x_1 \cdot x_2 \cdot \ldots \cdot x_n}$$

or, in logarithmic form,

(A.3b) $$\log G = \frac{\Sigma \log x}{n}$$

A.4 **Harmonic Mean**

$$H = \frac{n}{\Sigma \frac{1}{x}}$$

A.5 **Median** – If the x's are arranged according to size and one counts starting with the smallest value, the median $\tilde{x}$ is given by the $\frac{n+1}{2}$nd. Thus, when n is *odd* the median is the middle value in the ordered set of data, and when n is *even* it is the mean of the two middle values.

A.6 **Quartiles** – If the x's are arranged according to size and one counts starting with the smallest value, the jth quartile Q_j, $j = 1, 2$, or 3, is given by the $\frac{j(n+1)}{4}$th. Note that it may be necessary to interpolate between successive values.

A.7 **Deciles** – If the x's are arranged according to size and one counts starting with the smallest value, the jth decile D_j, $j = 1, 2, \ldots$, or 9, is given by the $\frac{j(n+1)}{10}$th. Note that it may be necessary to interpolate between successive values.

A.8 **Percentiles** – If the x's are arranged according to size and one counts starting with the smallest value, the jth percentile P_j, $j = 1, 2, \ldots$, or 99, is given by the $\frac{j(n+1)}{100}$th. Note that it may be necessary to interpolate between successive values.

A.9 **Mean Deviation**

$$MD = \frac{\Sigma |x - \bar{x}|}{n} \quad \text{or} \quad \frac{\Sigma |x - \tilde{x}|}{n}$$

where $\bar{x}$ is the mean and $\tilde{x}$ the median of the x's.

A.10 Standard Deviation

(A.10a)
$$s = \sqrt{\frac{\Sigma(x - \bar{x})^2}{n - 1}}$$

Equivalent computing formulas are

(A.10b)
$$s = \sqrt{\frac{\Sigma x^2 - n \cdot \bar{x}^2}{n - 1}}$$

(A.10c)
$$s = \sqrt{\frac{n(\Sigma x^2) - (\Sigma x)^2}{n(n - 1)}}$$

A.11 Variance – The variance is the square of the standard deviation. *See* **A.10.**

A.12 Interquartile Range

$$Q_3 - Q_1$$

where Q_1 and Q_3 are the first and third quartiles.

A.13 Quartile Deviation

$$\frac{Q_3 - Q_1}{2}$$

where Q_1 and Q_3 are the first and third quartiles; it is also called the *semi-interquartile range.*

A.14 Range

$$u_n - u_1$$

where u_n is the largest of the x's and u_1 is the smallest.

A.15 Coefficient of Variation

$$V = \frac{s}{\bar{x}} \cdot 100$$

where $\bar{x}$ and s are, respectively, the mean and the standard deviation of the x's.

A.16 Coefficient of Quartile Variation

$$V_Q = \frac{Q_3 - Q_1}{Q_3 + Q_1} \cdot 100$$

where Q_1 and Q_3 are the first and third quartiles of the x's.

A.17 Moments – The rth moment *about the origin* is

(A.17a) $$m_r' = \frac{\Sigma x^r}{n}$$

and the rth moment *about the mean* is

(A.17b) $$m_r = \frac{\Sigma (x - \bar{x})^r}{n}$$

The calculation of moments about the mean may be simplified by expanding $(x - \bar{x})^r$ according to the binomial theorem; moments about the mean can thus be expressed in terms of moments about the origin. *See also* E.8

A.18 Alpha-Three

$$\alpha_3 = \frac{m_3}{(m_2)^{3/2}}$$

where m_2 and m_3 are the second and third moments of the x's about their mean.

A.19 Alpha-Four

$$\alpha_4 = \frac{m_4}{(m_2)^2}$$

where m_2 and m_4 are the second and fourth moments of the x's about their mean.

A.20 Pearsonian Coefficient of Skewness

$$SK = \frac{3(\bar{x} - \tilde{x})}{s}$$

where $\bar{x}$, $\tilde{x}$, and s are, respectively, the mean, the median, and the standard deviation of the x's.

B. THE DESCRIPTION OF GROUPED DATA

The formulas of this section apply to data grouped into a frequency distribution having the class marks $x_1, x_2, \ldots,$ and x_k, and the corresponding class frequencies $f_1, f_2, \ldots,$ and f_k;

$$\sum_{i=1}^{k} f_i = n$$

is the total number of observations. The appearance of summations is simplified by omitting subscripts and limits.

B.1 Mean (Arithmetic Mean)

(B.1a) $$\bar{x} = \frac{x_1 f_1 + x_2 f_2 + \ldots + x_k f_k}{n} = \frac{\Sigma x \cdot f}{n}$$

or, with *coding*,

(B.1b) $$\bar{x} = x_0 + c \cdot \frac{\Sigma u \cdot f}{n}$$

where c is the class interval of the distribution and the u's are coded class marks obtained by replacing the x's with the integers $\ldots, -3, -2, -1, 0, 1, 2, 3, \ldots$ (with 0 corresponding to the class mark x_0 in the original scale).

B.2 Geometric Mean

(B.2a) $$G = \sqrt[n]{x_1^{f_1} \cdot x_2^{f_2} \cdot \ldots \cdot x_k^{f_k}}$$

or, in logarithmic form,

(B.2b) $$\log G = \frac{\Sigma(\log x) \cdot f}{n}$$

B.3 Median

(B.3a) $$\tilde{x} = L + c \cdot \frac{i}{f}$$

or

(B.3b) $$\tilde{x} = U - c \cdot \frac{i'}{f}$$

The first formula is used when one counts $n/2$ cases starting at the bottom of the distribution, namely, starting with the smallest values; L, c, and f are, respectively, the lower boundary, the interval, and the frequency of the class containing the median; i is the number of cases one still has to count after reaching L. The second formula is used when one counts $n/2$ cases starting at the top of the distribution, namely, starting with the largest values; U is the upper boundary of the class containing the median and i' is the number of cases one still has to count after reaching U.

B.4 **Quartiles** – The jth quartile Q_j is obtained with the use of formula (**B.3a**) counting $\frac{j \cdot n}{4}$ cases starting at the bottom of the distribution, or with the use of formula (**B.3b**) counting $n - \frac{j \cdot n}{4}$ cases starting at the top of the distribution.

B.5 **Deciles** – The jth decile D_j is obtained with the use of formula (**B.3a**) counting $\frac{j \cdot n}{10}$ cases starting at the bottom of the distribution, or with the use of formula (**B.3b**) counting $n - \frac{j \cdot n}{10}$ cases starting at the top of the distribution.

B.6 **Percentiles** – The jth percentile P_j is obtained with the use of formula (**B.3a**) counting $\frac{j \cdot n}{100}$ cases starting at the bottom of the distribution, or with the use of formula (**B.3b**) counting $n - \frac{j \cdot n}{100}$ cases starting at the top of the distribution.

B.7 **Mean Deviation**

$$MD = \frac{\Sigma |x - \bar{x}| \cdot f}{n} \quad \text{or} \quad \frac{\Sigma |x - \tilde{x}| \cdot f}{n}$$

where $\bar{x}$ is the mean and $\tilde{x}$ the median of the distribution.

B.8 Standard Deviation

(B.8a) $$s = \sqrt{\frac{\Sigma(x - \bar{x})^2 \cdot f}{n - 1}}$$

Equivalent computing formulas are

(B.8b) $$s = \sqrt{\frac{\Sigma x^2 f - n \cdot \bar{x}^2}{n - 1}}$$

(B.8c) $$s = \sqrt{\frac{n(\Sigma x^2 f) - (\Sigma xf)^2}{n(n - 1)}}$$

or, with *coding*,

(B.8d) $$s = c\sqrt{\frac{n(\Sigma u^2 f) - (\Sigma uf)^2}{n(n - 1)}}$$

where c is the class interval of the distribution and the u's are coded class marks obtained by replacing the x's with the integers $\ldots, -3, -2, -1, 0, 1, 2, 3, \ldots$.

B.9 Variance – The variance is the square of the standard deviation. *See* **B.8**

B.10 Interquartile Range – *See* formula A.12

B.11 Quartile Deviation – *See* formula A.13

B.12 Coefficient of Variation – *See* formula A.15

B.13 Coefficient of Quartile Variation – *See* formula A.16

B.14 Moments – The rth moment *about the origin* is

(B.14a) $$m_r' = \frac{\Sigma x^r f}{n}$$

and the rth moment *about the mean* is

(B.14b) $$m_r = \frac{\Sigma(x - \bar{x})^r f}{n}$$

The calculation of moments about the mean may be simplified by expanding $(x - \bar{x})^r$ according to the binomial theorem;

moments about the mean can thus be expressed in terms of moments about the origin. *See also* E.8

B.15 **Alpha-Three** – *See* formula A.18

B.16 **Alpha-Four** – *See* formula A.19

B.17 **Pearsonian Coefficient of Skewness** – *See* formula A.20

C. THE DESCRIPTION OF FINITE POPULATIONS

The formulas of this section apply to finite populations consisting of the elements (numbers) a_1, a_2, ..., and a_N. The appearance of summations is simplified by omitting subscripts and limits.

C.1 **Mean**

$$\mu = \frac{\Sigma a}{N}$$

C.2 **Standard Deviation**

(C.2a) $$\sigma = \sqrt{\frac{\Sigma (a - \mu)^2}{N}}$$

or

(C.2b) $$S = \sqrt{\frac{\Sigma (a - \mu)^2}{N - 1}}$$

Computing formulas are analogous to those of formula A.10.

C.3 **Variance** – The variance is the square of the standard devition. *See* formula C.2

C.4 **Moments** – The rth moment *about the origin* is

(C.4a) $$\mu_r' = \frac{\Sigma a^r}{N}$$

and the rth moment *about the mean* is

(C.4b) $$\mu_r = \frac{\Sigma (a - \mu)^r}{N}$$

The calculation of moments about the mean may be simplified by expanding $(a - \mu)^r$ according to the binomial theorem; moments about the mean can thus be expressed in terms of moments about the origin. *See also* E.8

D. PROBABILITY

The formulas of this section apply to events A, B, A_1, A_2, ..., in a given finite or countably infinite sample space S. The probability of event A is denoted $P(A)$.

D.1 **Postulates of Probability** – For a *finite* sample space S

(D.1a) $P(A) \geq 0$ for any event A

(D.1b) $P(S) = 1$

(D.1c) If A and B are mutually exclusive events,

$$P(A \cup B) = P(A) + P(B)$$

For a *countably infinite* sample space S, **(D.1c)** is replaced by

(D.1d) If A_1, A_2, A_3, ..., is a sequence of mutually exclusive events,

$$P\left(\bigcup_{i=1}^{\infty} A_i\right) = \sum_{i=1}^{\infty} P(A_i)$$

Here

$$\bigcup_{i=1}^{\infty} A_i$$

denotes the union of the events A_1, A_2, A_3, ..., and

$$\sum_{i=1}^{\infty} P(A_i)$$

denotes the sum of their respective probabilities.

D.2 **Conditional Probability** – If $P(B) \neq 0$, the conditional probability of A relative to B is given by

$$P(A|B) = \frac{P(A \cap B)}{P(B)}$$

D.3 **Independent Events** – Two events A and B are independent if and only if

$$P(A \cap B) = P(A) \cdot P(B)$$

Alternately, two events A and B are independent if and only if

$$P(A|B) = P(A) \quad \text{or} \quad P(B) = 0$$

More generally, k events $A_1, A_2, \ldots,$ and A_k are independent if and only if for any subset of r of these events ($r = 2, 3, \ldots,$ and k) the probability of their intersection equals the product of their respective probabilities.

D.4 **Probability of the Empty Set** $\emptyset$

$$P(\emptyset) = 0$$

D.5 **Probability of Complement** – If A' is the complement of event A relative to the sample space S, then

$$P(A') = 1 - P(A)$$

D.6 **Generalization of Postulate (D.1c)** – If $A_1, A_2, \ldots,$ and A_k are (pairwise) mutually exclusive events,

$$P(A_1 \cup A_2 \cup \ldots \cup A_k) = \sum_{i=1}^{k} P(A_i)$$

D.7 **General Rule of Addition** – For any two events A_1 and A_2

(D.7a) $$P(A_1 \cup A_2) = P(A_1) + P(A_2) - P(A_1 \cap A_2)$$

For any three events A_1, A_2, and A_3,

(D.7b) $$P(A_1 \cup A_2 \cup A_3) = P(A_1) + P(A_2) + P(A_3) - P(A_1 \cap A_2) - P(A_1 \cap A_3) - P(A_2 \cap A_3) + P(A_1 \cap A_2 \cap A_3)$$

More generally, there exists a corresponding formula for the union of any k events.

D.8 **General Rule of Multiplication** – For any two events A_1 and A_2

(D.8a)
$$P(A_1 \cap A_2) = P(A_1) \cdot P(A_2|A_1) \quad \text{provided } P(A_1) \neq 0$$
$$= P(A_2) \cdot P(A_1|A_2) \quad \text{provided } P(A_2) \neq 0$$

For any three events A_1, A_2, and A_3,

(D.8b)
$$P(A_1 \cap A_2 \cap A_3) = P(A_1) \cdot P(A_2|A_1) \cdot P(A_3|A_1 \cap A_2)$$
$$\text{provided } P(A_1) \neq 0 \quad \text{and} \quad P(A_1 \cap A_2) \neq 0$$

More generally, there exists a corresponding formula for the intersection of any k events.

D.9 **Special Rule for Equiprobable Events** – If an experiment has n possible equiprobable outcomes among which s are labeled "success," the probability of a "success" is s/n.

D.10 **Rule of Elimination** – If B_1, B_2, ..., and B_k are mutually exclusive events for which $B_1 \cup B_2 \cup \ldots \cup B_k = S$ and $P(B_i) \neq 0$ for $i = 1, 2, \ldots$, and k, then for any event A

$$P(A) = \sum_{i=1}^{k} P(B_i) \cdot P(A|B_i)$$

D.11 **Bayes' Theorem** – If B_1, B_2, ..., and B_k are mutually exclusive events for which $B_1 \cup B_2 \cup \ldots \cup B_k = S$ and $P(B_i) \neq 0$ for $i = 1, 2, \ldots$, and k, then for any event A

(D.11a)
$$P(B_r|A) = \frac{P(B_r) \cdot P(A|B_r)}{\sum_{i=1}^{k} P(B_i) \cdot P(A|B_i)}$$

for $r = 1, 2, \ldots$, or k.

(In the continuous case, the formula becomes

(D.11b)
$$f(y|x) = \frac{f(x|y)f(y)}{\int f(x|y)f(y)dy}$$

where $f(y)$ is the *a priori* (or *prior*) *density* of the random variable Y, and $f(y|x)$ is the *a posteriori* (or *posterior*) *density* of Y for the given value of X. In problems of Bayesian inference, Y is generally a parameter concerning which the inference is to be made.)

D.12 **Odds** – The odds for the occurrence of event A are "$P(A)$ to $1 - P(A)$" or any two positive numbers which are in the same ratio; generally, odds are expressed as ratios of positive integers.

D.13 **Mathematical Expectation** – If $A_1, A_2, \ldots,$ and A_k are the k mutually exclusive events of obtaining the amounts $a_1, a_2, \ldots,$ and a_k, then the expected amount, or the mathematical expectation, is

$$\sum_{i=1}^{k} a_i \cdot P(A_i)$$

E. PROBABILITY DISTRIBUTIONS

Random variables with values x (y, z, x_1, x_2, etc.) are denoted X (Y, Z, X_1, X_2, etc.); an alternate notation is to use boldface type, that is, denote the corresponding random variables $\mathbf{x}$ ($\mathbf{y}$, $\mathbf{z}$, $\mathbf{x}_1$, $\mathbf{x}_2$, etc.).

E.1 **Probability Function** — A function with values $f(x)$ defined over the domain of a random variable X, for which

(E.1a) $f(x) \geq 0$ for each x within the domain of X

(E.1b) $\Sigma f(x) = 1$, where the summation extends over the domain of X

(E.1c) $f(x)$ is the probability that the random variable has the value x.

E.2 **Probability Density Function** — A function with values $f(x)$ defined for $-\infty < x < \infty$, for which

(E.2a) $f(x) \geq 0$ for $-\infty < x < \infty$

(E.2b) $$\int_{-\infty}^{\infty} f(x)dx = 1$$

(E.2c) $\int_{a}^{b} f(x)dx$ is the probability that a corresponding random variable X assumes a value on the interval from a to b ($a \leq b$).

E.3 **Distribution Function** — For a discrete random variable X, the function with values $F(t)$ for which

(E.3a) $$F(t) = \sum_{x \leq t} f(x) \quad \text{for } -\infty < t < \infty$$

where $f(x)$ is a value of the probability function of X. For a continuous random variable X, the function with values $F(t)$ for which

(E.3b) $$F(t) = \int_{-\infty}^{t} f(x)dx \quad \text{for } -\infty < t < \infty$$

where $f(x)$ is a value of the probability density function of X. Note that

(E.3c) $$F'(t) = f(t)$$

where this derivative exists.

E.4 **Expected Value** – For a discrete random variable X,

(E.4a) $$E(X) = \Sigma x \cdot f(x),$$

where the summation extends over the domain of X; more generally,

(E.4b) $$E[g(X)] = \Sigma g(x) \cdot f(x)$$

where the summation extends over the domain of X. For a continuous random variable X,

(E.4c) $$E(X) = \int_{-\infty}^{\infty} x \cdot f(x) dx$$

and more generally

(E.4d) $$E[g(X)] = \int_{-\infty}^{\infty} g(x) \cdot f(x) dx$$

E.5 **Mean**

$$\mu = E(X)$$

E.6 **Standard Deviation**

$$\sigma = \sqrt{E[(X - \mu)^2]}$$

E.7 **Variance**

$$\sigma^2 = E[(X - \mu)^2]$$

E.8 **Moments** – The rth moment *about the origin* is

(E.8a) $$\mu_r' = E(X^r)$$

and the rth moment *about the mean* is

(E.8b) $$\mu_r = E[(X - \mu)^r]$$

The determination of moments about the mean may be simplified by expanding $(X - \mu)^r$ according to the binomial theorem; moments about the mean can thus be expressed in terms of moments about the origin. For example,

(E.8c) $$\mu_2 = \mu_2' - (\mu_1')^2 = \mu_2' - \mu^2$$

(E.8d) $$\mu_3 = \mu_3' - 3\mu_2'\mu + 2\mu^3$$

(E.8e) $$\mu_4 = \mu_4' - 4\mu_3'\mu + 6\mu_2'\mu^2 - 3\mu^4$$

E.9 **Moment Generating Function** – For a given random variable X, the function with values

$$M(\theta) = E(e^{\theta X}) = \sum_{k=0}^{\infty} \frac{\theta^k}{k!} \cdot \mu_k'$$

E.10 **Characteristic Function** – For a given random variable X, the function with values

$$\emptyset(t) = E(e^{itX}) = \sum_{k=0}^{\infty} \frac{(it)^k}{k!} \cdot \mu_k'$$

E.11 **Cumulant** – The rth cumulant κ_r is the coefficient of $\theta^r/r!$ in the Maclaurin's series of the logarithm (to the base e) of the moment generating function of a random variable.

E.12 **Factorial Moments** – The rth factorial moment is

$$\mu_{(r)}' = E[X(X - 1)(X - 2) \cdot \ldots \cdot (X - r + 1)]$$

E.13 **Alpha-Three**

$$\alpha_3 = \frac{\mu_3}{(\mu_2)^{3/2}}$$

E.14 **Alpha-Four**

$$\alpha_4 = \frac{\mu_4}{(\mu_2)^2}$$

E.15 **Joint Probability Function** – A function with values $f(x_1, x_2, \ldots, x_k)$, defined over the domain of a set of random variables X_1, X_2, ..., and X_k, for which

(E.15a) $f(x_1, x_2, \ldots, x_k) \geq 0$ for each set of values within the domain of the random variables;

(E.15b) $\sum_{x_1} \sum_{x_2} \cdots \sum_{x_k} f(x_1, x_2, \ldots, x_k) = 1$, where the summations extend over the respective domains of the random variables;

(E.15c) $f(x_1, x_2, \ldots, x_k)$ is the probability that the k random variables have the values x_1, x_2, ..., and x_k, respectively.

E.16 **Joint Density Function** – A function with values $f(x_1, x_2, \ldots, x_k)$, defined for $-\infty < x_i < \infty$ for $i = 1, 2, \ldots$, and k, for which

(E.16a) $f(x_1, x_2, \ldots, x_k) \geq 0$ for $-\infty < x_i < \infty$, $i = 1, 2, \ldots$, and k

(E.16b)
$$\int_{-\infty}^{\infty} \int_{-\infty}^{\infty} \cdots \int_{-\infty}^{\infty} f(x_1, x_2, \ldots, x_k) dx_1 dx_2 \ldots dx_k = 1$$

(E.16c)
$$\int_{a_1}^{b_1} \int_{a_2}^{b_2} \cdots \int_{a_k}^{b_k} f(x_1, x_2, \ldots, x_k) dx_1 dx_2 \ldots dx_k$$

is the probability that random variable X_1 assumes a value between a_1 and b_1, random variable X_2 assumes a value between a_2 and b_2, ..., and random variable X_k assumes a value between a_k and b_k.

E.17 **Joint Distribution Function** – If $f(x_1, x_2, \ldots, x_k)$ is a value of the joint probability function of k random variables X_1, X_2, ..., and X_k, their joint distribution function is given by

(E.17a)
$$F(t_1, t_2, \ldots, t_k) = \sum_{x_1 \leq t_1} \sum_{x_2 \leq t_2} \cdots \sum_{x_k \leq t_k} f(x_1, x_2, \ldots, x_k)$$

for $-\infty < t_i < \infty$, $i = 1, 2, \ldots, k$.

If $f(x_1, x_2, \ldots, x_k)$ is a value of the joint probability density of k random variables $X_1, X_2, \ldots,$ and X_k, their joint distribution function is given by

(E.17b)
$$F(t_1, t_2, \ldots, t_k) = \int_{-\infty}^{t_1} \int_{-\infty}^{t_2} \cdots \int_{-\infty}^{t_k} f(x_1, x_2, \ldots, x_k)\,dx_1 dx_2 \ldots dx_k$$

for $-\infty < t_i < \infty$, $i = 1, 2, \ldots, k$.

E.18 **Marginal Distribution** – If $f(x_1, x_2, \ldots, x_k)$ is a value of the joint probability function of k random variables $X_1, X_2, \ldots,$ and X_k, the marginal distribution of X_i ($i = 1, 2, \ldots,$ or k) is given by

(E.18a)
$$f_i(x_i) = \sum_{x_1} \cdots \sum_{x_{i-1}} \sum_{x_{i+1}} \cdots \sum_{x_k} f(x_1, x_2, \ldots, x_k)$$

where the summations extend over the respective domains of the random variables.

If $f(x_1, x_2, \ldots, x_k)$ is a value of the joint probability density of k random variables $X_1, X_2, \ldots,$ and X_k, the marginal distribution (or density) of X_i ($i = 1, 2, \ldots,$ or k) is given by

(E.18b)
$$f_i(x_i) = \int_{-\infty}^{\infty} \cdots \int_{-\infty}^{\infty} \int_{-\infty}^{\infty} \cdots \int_{-\infty}^{\infty} f(x_1, \ldots, x_k)\,dx_1 \ldots dx_{i-1} dx_{i+1} \ldots dx_k$$

E.19 **Conditional Distribution** – If $f(x_1, x_2)$ is a value of the joint probability function (or probability density) of two random variables X_1 and X_2, the conditional distribution (probability function or probability density) of X_1 *when* X_2 *has the value* x_2 is given by

$$f(x_1|x_2) = \frac{f(x_1, x_2)}{f_2(x_2)}$$

E.20 **Product Moments** – The rth and sth product moment of two random variables X_1 and X_2 *about their origins* is

(E.20a) $$\mu'_{rs} = E(X_1^r X_2^s)$$

and the rth and sth product moment of the two random variables *about their means* is

(E.20b) $$\mu_{rs} = E[(X_1 - \mu_1)^r (X_2 - \mu_2)^s]$$

where $\mu_1 = E(X_1)$ and $\mu_2 = E(X_2)$.

E.21 **Covariance** – The covariance of two random variables X_1 and X_2 is

$$\sigma_{12} = \mu_{11} = E[(X_1 - \mu_1)(X_2 - \mu_2)]$$

where $\mu_1 = E(X_1)$ and $\mu_2 = E(X_2)$; the covariance is also denoted Cov (X_1, X_2) or $C(X_1, X_2)$.

E.22 **Correlation Coefficient** – For two random variables X_1 and X_2

$$\rho = \frac{\mu_{11}}{\sqrt{\mu_{20} \cdot \mu_{02}}}$$

where μ_{11}, μ_{20}, and μ_{02} are as defined in E.20.

E.23 **Binomial Distribution** – The distribution of a random variable having the probability function

(E.23a) $$f(x) = \binom{n}{x} \theta^x (1 - \theta)^{n-x} \quad \text{for } x = 0, 1, 2, \ldots, n$$

The parameters of this distribution are n and θ, the mean is

(E.23b) $$\mu = n\theta$$

and the variance is

(E.23c) $$\sigma^2 = n\theta(1 - \theta)$$

E.24 **Discrete Uniform Distribution** – The distribution of a random variable having the probability function

(E.24a) $$f(x) = 1/n \quad \text{for } x = x_1, x_2, \ldots, x_n$$

When $x_i = i$ for $i = 1, 2, \ldots$, and n, the mean is

(E.24b) $$\mu = \frac{n+1}{2}$$

and the variance is

(E.24c) $$\sigma^2 = \frac{n^2 - 1}{12}$$

E.25 **Geometric Distribution** – The distribution of a random variable having the probability function

(E.25a) $$f(x) = \theta(1 - \theta)^{x-1} \quad \text{for } x = 1, 2, 3, \ldots.$$

The parameter of this distribution is $\theta > 0$, the mean is

(E.25b) $$\mu = 1/\theta$$

and the variance is

(E.25c) $$\sigma^2 = \frac{1 - \theta}{\theta^2}$$

It is a special case of the *negative binomial distribution* with $k = 1$.

E.26 **Hypergeometric Distribution** – The distribution of a random variable having the probability function

(E.26a) $$f(x) = \frac{\binom{a}{x}\binom{b}{n-x}}{\binom{a+b}{n}} \quad \text{for } x = 0, 1, 2, \ldots, n$$

The parameters of this distribution are the positive integers

a, b, and n, or n, $N = a + b$, and $\theta = \frac{a}{a+b}$; the mean is

(E.26b) $$\mu = \frac{na}{a+b} = n\theta$$

and the variance is

(E.26c) $$\sigma^2 = \frac{nab(a+b-n)}{(a+b)^2(a+b-1)} = n\theta(1-\theta)\frac{N-n}{N-1}$$

E.27 **Negative Binomial Distribution** – The distribution of a random variable having the probability function

(E.27a) $$f(x) = \binom{x-1}{k-1}\theta^k(1-\theta)^{x-k} \quad \text{for } x = k, k+1, k+2, \ldots\ldots$$

The parameters of this distribution are k and $\theta > 0$, the mean is

(E.27b) $$\mu = k/\theta$$

and the variance is

(E.27c) $$\sigma^2 = \frac{k}{\theta}\left(\frac{1}{\theta} - 1\right)$$

E.28 **Poisson Distribution** – The distribution of a random variable having the probability function

(E.28a) $$f(x) = \frac{\lambda^x e^{-\lambda}}{x!} \quad \text{for } x = 0, 1, 2, \ldots\ldots$$

The parameter of this distribution is $\lambda > 0$, the mean is

(E.28b) $$\mu = \lambda$$

and the variance is

(E.28c) $$\sigma^2 = \lambda$$

E.29 **Multinomial Distribution** – The joint distribution of k random variables $X_1, X_2, \ldots,$ and X_k, having the joint probability function

(E.29a)
$$f(x_1, x_2, \ldots, x_k) = \frac{n!}{x_1!x_2!\cdot\ldots\cdot x_k!}\theta_1^{x_1}\theta_2^{x_2}\cdot\ldots\cdot\theta_k^{x_k}$$

for $x_i = 0, 1, 2, \ldots, n$ $(i = 1, 2, \ldots, k)$, subject to the restrictions

$$\sum_{i=1}^{k} x_i = n \quad \text{and} \quad \sum_{i=1}^{k}\theta_i = 1.$$

The parameters of this distribution are $n, \theta_1, \theta_2, \ldots,$ and θ_k, the mean of X_i is

(E.29b)
$$\mu_i = n\theta_i \quad \text{for } i = 1, 2, \ldots, k$$

the variance of X_i is

(E.29c)
$$\sigma_i^2 = n\theta_i(1 - \theta_i) \quad \text{for } i = 1, 2, \ldots, k$$

and the covariance of X_i and X_j is

(E.29d)
$$\sigma_{ij} = -n\theta_i\theta_j \quad \text{for } i \neq j = 1, 2, \ldots, k$$

E.30 **Beta Distribution** – The distribution of a random variable having the probability density

(E.30a)
$$f(x) = \begin{cases} \dfrac{\Gamma(\alpha+\beta)}{\Gamma(\alpha)\Gamma(\beta)}x^{\alpha-1}(1-x)^{\beta-1} & \text{for } 0 < x < 1 \\ 0 & \text{elsewhere} \end{cases}$$

where $\Gamma(\alpha)$ is as defined in E.35. The parameters of this distribution are $\alpha > 0$ and $\beta > 0$, the mean is

(E.30b)
$$\mu = \frac{\alpha}{\alpha + \beta}$$

and the variance is

(E.30c)
$$\sigma^2 = \frac{\alpha\beta}{(\alpha + \beta)^2(\alpha + \beta + 1)}$$

E.31 **Cauchy Distribution** — The distribution of a random variable having the probability density

$$f(x) = \frac{1}{\pi} \cdot \frac{a}{a^2 + x^2} \quad \text{for } -\infty < x < \infty$$

The parameter of this distribution is $a > 0$; its moments do not exist.

E.32 **Chi-Square Distribution** — The distribution of a random variable having the probability density

(E.32a) $$f(x) = \begin{cases} \dfrac{1}{2^{\nu/2}\Gamma(\nu/2)} x^{(\nu-2)/2} e^{-x/2} & \text{for } x > 0 \\ 0 & \text{elsewhere} \end{cases}$$

where $\Gamma(\nu/2)$ is as defined in E.35. The parameter of this distribution is ν, the number of degrees of freedom, the mean is

(E.32b) $$\mu = \nu$$

and the variance is

(E.32c) $$\sigma^2 = 2\nu$$

The chi-square distribution is a special case of the gamma distribution, E.35, with $\alpha = \nu/2$ and $\beta = 2$.

E.33 **Exponential Distribution** — The distribution of a random variable having the probability density

(E.33a) $$f(x) = \begin{cases} \dfrac{1}{\theta} e^{-x/\theta} & \text{for } x > 0 \\ 0 & \text{elsewhere} \end{cases}$$

The parameter of this distribution is $\theta > 0$, the mean is

(E.33b) $$\mu = \theta$$

and the variance is

(E.33c) $$\sigma^2 = \theta^2$$

The exponential distribution is a special case of the gamma distribution, E.35, with $\alpha = 1$ and $\beta = \theta$.

E.34 ***F* Distribution** – The distribution of a random variable having the probability density

(E.34a) $$f(x) = \begin{cases} \dfrac{\Gamma\left(\dfrac{\nu_1 + \nu_2}{2}\right)}{\Gamma\left(\dfrac{\nu_1}{2}\right)\Gamma\left(\dfrac{\nu_2}{2}\right)} (\nu_1/\nu_2)^{\nu_1/2} x^{\nu_1/2-1}\left(1 + \dfrac{\nu_1}{\nu_2}x\right)^{-(\nu_1+\nu_2)/2} & \text{for } x > 0 \\ 0 & \text{elsewhere} \end{cases}$$

where the values of Γ are as defined in E.35. The parameters of this distribution are ν_1 and ν_2, its respective degrees of freedom, the mean is

(E.34b) $$\mu = \frac{\nu_2}{\nu_2 - 2}$$

and the variance is

(E.34c) $$\sigma^2 = \frac{2\nu_2^2(\nu_1 + \nu_2 - 2)}{\nu_1(\nu_2 - 2)^2(\nu_2 - 4)}$$

E.35 **Gamma Distribution** – The distribution of a random variable having the probability density

(E.35a) $$f(x) = \begin{cases} \dfrac{1}{\beta^{\alpha}\Gamma(\alpha)} x^{\alpha-1}e^{-x/\beta} & \text{for } x > 0 \\ 0 & \text{elsewhere} \end{cases}$$

where

(E.35b) $$\Gamma(\alpha) = \int_0^\infty x^{\alpha-1} e^{-x} dx$$

and, in particular, $\Gamma(\alpha) = (\alpha - 1)!$ when α is a positive integer. The parameters of this distribution are $\alpha > 0$ and $\beta > 0$, the mean is

(E.35c) $$\mu = \beta\alpha$$

and the variance is

(E.35d) $$\sigma^2 = \beta^2\alpha$$

The gamma distribution is also referred to as the *Erlang distribution of order* α.

E.36 **Normal Distribution** – The distribution of a random variable having the probability density

(E.36a) $$f(x) = \frac{1}{\sigma\sqrt{2\pi}} e^{-\frac{1}{2}\left(\frac{x-\mu}{\sigma}\right)^2} \quad \text{for } -\infty < x < \infty$$

The parameters of this distribution are μ and σ, its mean and standard deviation. When $\mu = 0$, and $\sigma = 1$, the distribution is referred to as the *standard normal distribution.*

E.37 ***t* Distribution** – The distribution of a random variable having the probability density

(E.37a) $$f(x) = \frac{\Gamma\left(\frac{\nu+1}{2}\right)}{\sqrt{\pi\nu}\,\Gamma(\nu/2)}\left(1 + \frac{x^2}{\nu}\right)^{-(\nu+1)/2} \quad \text{for } -\infty < x < \infty$$

The parameter of this distribution is ν, the number of degrees of freedom; the mean is 0 and the variance is

(E.37b) $$\sigma^2 = \frac{\nu}{\nu - 2}$$

E.38 **Uniform Distribution** — The distribution of a random variable having the probability density

(E.38a)
$$f(x) = \begin{cases} \dfrac{1}{\beta - \alpha} & \text{for } \alpha < x < \beta \\ 0 & \text{elsewhere} \end{cases}$$

The parameters of this distribution are α and β, the mean is

(E.38b)
$$\mu = \frac{\alpha + \beta}{2}$$

and the variance is

(E.38c)
$$\sigma^2 = \frac{(\beta - \alpha)^2}{12}$$

E.39 **Weibull Distribution** — The distribution of a random variable having the probability density

(E.39a)
$$f(x) = \begin{cases} \alpha\beta x^{\beta-1} e^{-\alpha x^{\beta}} & \text{for } x > 0 \\ 0 & \text{elsewhere} \end{cases}$$

The parameters of this distribution are $\alpha > 0$ and $\beta > 0$, the mean is

(E.39b)
$$\mu = \alpha^{-1/\beta}\Gamma\left(1 + \frac{1}{\beta}\right)$$

and the variance is

(E.39c)
$$\sigma^2 = \alpha^{-2/\beta}\left\{\Gamma\left(1 + \frac{2}{\beta}\right) - \left[\Gamma\left(1 + \frac{1}{\beta}\right)\right]^2\right\}$$

where the values of Γ are as defined in E.35.

E.40 **Rayleigh Distribution** — The distribution of a random variable having the probability density

$$f(x) = \begin{cases} 2\alpha x e^{-\alpha x^2} & \text{for } x > 0 \\ 0 & \text{elsewhere} \end{cases}$$

This distribution is a special case of the Weibull distribution, E.39, with $\beta = 2$.

E.41 **Bivariate Normal Distribution** — The joint distribution of a pair of random variables X_1 and X_2 having the joint probability density

$$f(x_1,x_2) = \frac{e^{-\left[\left(\frac{x_1-\mu_1}{\sigma_1}\right)^2 - 2\rho\left(\frac{x_1-\mu_1}{\sigma_1}\right)\left(\frac{x_2-\mu_2}{\sigma_2}\right) + \left(\frac{x_2-\mu_2}{\sigma_2}\right)^2\right] \Big/ 2(1-\rho^2)}}{2\pi\sigma_1\sigma_2\sqrt{1-\rho^2}}$$

for $-\infty < x_1 < \infty$ and $-\infty < x_2 < \infty$. The parameters of this distribution are μ_1, μ_2, σ_1, σ_2, and ρ, which are, respectively, the means of the two random variables, their standard deviations, and the correlation coefficient as defined in E.22.

F. FORMULAS FOR INFERENCES CONCERNING MEANS AND STANDARD DEVIATIONS

Unless specified otherwise, the formulas of this section apply to a random sample of size n from an infinite population with the mean μ and the standard deviation σ, or to two independent random samples of size n_1 and n_2 from two infinite populations with the means μ_1 and μ_2 and the standard deviations σ_1 and σ_2. The means of the samples are denoted $\bar{x}$, $\bar{x}_1$, and $\bar{x}_2$, and their standard deviations are denoted s, s_1, and s_2; the samples are regarded as large when the sample size is 30 or more. For all confidence limits the degree of confidence is $1 - \alpha$, and for all tests the level of significance is α. The symbol z_α denotes the value for which the area to its right under the standard normal curve is equal to α; $t_{\alpha,\nu}$, $\chi^2_{\alpha,\nu}$, and F_{α,ν_1,ν_2} are defined correspondingly for the t distribution with ν degrees of freedom, the chi-square distribution with ν degrees of freedom, and the F distribution with ν_1 and ν_2 degrees of freedom.

F.1 Standard Error of the Mean

$$\sigma_{\bar{x}} = \frac{\sigma}{\sqrt{n}}$$

F.2 Standard Error of the Mean (Finite Population)

$$\sigma_{\bar{x}} = \frac{\sigma}{\sqrt{n}}\sqrt{\frac{N-n}{N-1}}$$

F.3 Standard Error of the Mean (Stratified Sample)—For stratified samples from a finite population of size N, divided into k strata of size N_1, N_2, . . . , and N_k,

$$\sigma_{\bar{x}} = \frac{1}{N}\sqrt{\sum_{i=1}^{k} N_i(N_i - n_i)\cdot\frac{S_i^2}{n_i}}$$

where n_i is the number of observations taken from the ith stratum and S_i is the standard deviation of the ith stratum as defined in C.2.

F.4 Standard Error of Difference Between Two Means

$$\sigma_{\bar{x}_1 - \bar{x}_2} = \sqrt{\frac{\sigma_1^2}{n_1} + \frac{\sigma_2^2}{n_2}}$$

F.5 **Standard Error of the Median** – For large samples from normal populations,

$$\sigma_{\tilde{x}} = \sqrt{\frac{\pi}{2}} \cdot \frac{\sigma}{\sqrt{n}}$$

F.6 **Standard Error of the Standard Deviation** – For large samples

$$\sigma_s = \frac{\sigma}{\sqrt{2n}}$$

F.7 **Standard Error of the Variance**

(F.7a)
$$\sigma_{s^2} = \sqrt{\frac{\mu_4}{n} - \frac{(n-3)\sigma^4}{n(n-1)}}$$

where μ_4 is the population's fourth moment about the mean; for samples from normal populations

(F.7b)
$$\sigma_{s^2} = \sigma^2 \sqrt{\frac{2}{n-1}}$$

F.8 **Maximum Error of Estimate of μ (Normal Population, σ Known)** – When $\bar{x}$ is used as an estimate of μ, one can assert with the probability $1 - \alpha$ that the error $E = |\bar{x} - \mu|$ will not exceed

(F.8a)
$$E = z_{\alpha/2} \cdot \frac{\sigma}{\sqrt{n}}$$

Correspondingly, the minimum sample size required to attain this precision is

(F.8b)
$$n = \left(\frac{z_{\alpha/2} \cdot \sigma}{E}\right)^2$$

F.9 **Maximum Error of Estimate of μ (Normal Population, σ Unknown)** – When $\bar{x}$ is used as an estimate of μ, one can assert with the probability $1 - \alpha$ that the error $E = |\bar{x} - \mu|$ will not exceed

$$E = t_{\alpha/2, n-1} \cdot \frac{s}{\sqrt{n}}$$

F.10 Confidence Limits for μ (Normal Population, σ Known)

$$\bar{x} \pm z_{\alpha/2} \cdot \frac{\sigma}{\sqrt{n}}$$

These limits are also used for large samples from populations which are not necessarily normal, and with s substituted for σ when σ is unknown.

F.11 Confidence Limits for μ (Normal Population, σ Unknown)

$$\bar{x} \pm t_{\alpha/2,n-1} \cdot \frac{s}{\sqrt{n}}$$

F.12 Confidence Limits for μ (Finite Population, σ Known) – For large samples from a finite population of size N,

$$\bar{x} \pm z_{\alpha/2} \cdot \frac{\sigma}{\sqrt{n}} \sqrt{\frac{N-n}{N-1}}$$

These large-sample confidence limits should be used when the sample constitutes 5 per cent or more of the population; when σ is unknown, they are also used with s substituted for σ.

F.13 Confidence Limits for $\mu_1 - \mu_2$ (Normal Populations, σ_1 and σ_2 Known)

$$\bar{x}_1 - \bar{x}_2 \pm z_{\alpha/2} \sqrt{\frac{\sigma_1^2}{n_1} + \frac{\sigma_2^2}{n_2}}$$

These confidence limits are also used for large samples from populations which are not necessarily normal, and with s_1 and s_2 substituted for σ_1 and σ_2 when σ_1 and σ_2 are unknown.

F.14 Confidence Limits for $\mu_1 - \mu_2$ (Normal Populations, $\sigma_1 = \sigma_2$ Unknown)

$$\bar{x}_1 - \bar{x}_2 \pm t_{\alpha/2,n_1+n_2-2} \cdot \frac{\sqrt{(n_1-1)s_1^2 + (n_2-1)s_2^2}}{\sqrt{\dfrac{n_1 n_2(n_1+n_2-2)}{n_1+n_2}}}$$

F.15 **Confidence Limits for σ (Normal Populations)**

$$\sqrt{\frac{(n-1)s^2}{\chi^2_{\alpha/2,n-1}}} \quad \text{and} \quad \sqrt{\frac{(n-1)s^2}{\chi^2_{1-\alpha/2,n-1}}}$$

F.16 **Confidence Limits for σ (Large Samples)**

$$\frac{s}{1+\frac{z_{\alpha/2}}{\sqrt{2n}}} \quad \text{and} \quad \frac{s}{1-\frac{z_{\alpha/2}}{\sqrt{2n}}}$$

F.17 **One-Sample z Test for μ (Normal Population, σ Known)** – A test of the null hypothesis $\mu = \mu_0$ based on the statistic

(F.17a)
$$z = \frac{\sqrt{n}\,(\bar{x} - \mu_0)}{\sigma}$$

The null hypothesis is rejected for $z < -z_\alpha$ when the alternative hypothesis is $\mu < \mu_0$, for $z > z_\alpha$ when the alternative hypothesis is $\mu > \mu_0$, and for $z < -z_{\alpha/2}$ or $z > z_{\alpha/2}$ when the alternative hypothesis is $\mu \neq \mu_0$. This test is also used for large samples from populations which are not necessarily normal and with s substituted for σ when σ is unknown. If the alternative hypothesis is one-sided and the probability of a Type II error at $\mu = \mu_1$ is to be β, the required sample size is given by

(F.17b)
$$n = \frac{\sigma^2(z_\alpha + z_\beta)^2}{(\mu_1 - \mu_0)^2}$$

For the two-tail test, $z_{\alpha/2}$ must be substituted in this formula for z_α.

F.18 **One-Sample t Test for μ (Normal Population, σ Unknown)** – A test of the null hypothesis $\mu = \mu_0$ based on the statistic

$$t = \frac{\sqrt{n}\,(\bar{x} - \mu_0)}{s}$$

The null hypothesis is rejected for $t < -t_{\alpha,n-1}$ when the alternative hypothesis is $\mu < \mu_0$, for $t > t_{\alpha,n-1}$ when the alternative hypothesis is $\mu > \mu_0$, and for $t < -t_{\alpha/2,n-1}$ or $t > t_{\alpha/2,n-1}$ when the alternative hypothesis is $\mu \neq \mu_0$.

F.19 **Two-Sample z Test for $\mu_1 - \mu_2$ (Normal Populations, σ_1 and σ_2 Known)**—A test of the null hypothesis $\mu_1 - \mu_2 = \delta$ based on the statistic

$$z = \frac{\bar{x}_1 - \bar{x}_2 - \delta}{\sqrt{\frac{\sigma_1^2}{n_1} + \frac{\sigma_2^2}{n_2}}}$$

The null hypothesis is rejected for $z < -z_\alpha$ when the alternative hypothesis is $\mu_1 - \mu_2 < \delta$, for $z > z_\alpha$ when the alternative hypothesis is $\mu_1 - \mu_2 > \delta$, and for $z < -z_{\alpha/2}$ or $z > z_{\alpha/2}$ when the alternative hypothesis is $\mu_1 - \mu_2 \neq \delta$. This test is also used for large samples from populations which are not necessarily normal, and with s_1 and s_2 substituted for σ_1 and σ_2 when σ_1 and σ_2 are unknown.

F.20 **Two-Sample t Test for $\mu_1 - \mu_2$ (Normal Populations, $\sigma_1 = \sigma_2$ Unknown)**—A test of the null hypothesis $\mu_1 - \mu_2 = \delta$ based on the statistic

$$t = \frac{\bar{x}_1 - \bar{x}_2 - \delta}{\sqrt{(n_1 - 1)s_1^2 + (n_2 - 1)s_2^2}} \cdot \sqrt{\frac{n_1 n_2 (n_1 + n_2 - 2)}{n_1 + n_2}}$$

The null hypothesis is rejected for $t < -t_{\alpha, n_1+n_2-2}$ when the alternative hypothesis is $\mu_1 - \mu_2 < \delta$, for $t > t_{\alpha, n_1+n_2-2}$ when the alternative hypothesis is $\mu_1 - \mu_2 > \delta$, and for $t < -t_{\alpha/2, n_1+n_2-2}$ or $t > t_{\alpha/2, n_1+n_2-2}$ when the alternative hypothesis is $\mu_1 - \mu_2 \neq \delta$.

F.21 **Behrens-Fisher Test (Normal Populations, σ_1 and σ_2 Unknown)**—A test of the null hypothesis $\mu_1 - \mu_2 = \delta$ which is used instead of the two-sample t test when it cannot be assumed that $\sigma_1 = \sigma_2$; it is based on the statistic

(F.21a) $$d = \frac{\bar{x}_1 - \bar{x}_2 - \delta}{\sqrt{\frac{s_1^2}{n_1} + \frac{s_2^2}{n_2}}}$$

The critical values of this test, obtained by fiducial arguments, are tabulated in the Fisher-Yates Tables, listed on page 194, for various values of n_1, n_2, α, and θ, where

(F.21b) $$\theta = \text{arc tan} \frac{s_1}{s_2} \sqrt{\frac{n_2}{n_1}}$$

The theory on which this test is based can also be used to construct fiducial limits for $\mu_1 - \mu_2$.

F.22 **One-Sample χ^2 Test for σ (Normal Population)**—A test of the null hypothesis $\sigma = \sigma_0$ based on the statistic

$$\chi^2 = \frac{(n-1)s^2}{\sigma_0^2}$$

The null hypothesis is rejected for $\chi^2 < \chi^2_{1-\alpha,n-1}$ when the alternative hypothesis is $\sigma < \sigma_0$, for $\chi^2 > \chi^2_{\alpha,n-1}$ when the alternative hypothesis is $\sigma > \sigma_0$, and for $\chi^2 < \chi^2_{1-\alpha/2,n-1}$ or $\chi^2 > \chi^2_{\alpha/2,n-1}$ when the alternative hypothesis is $\sigma \neq \sigma_0$.

F.23 **Large-Sample z Test for σ**—A large-sample test of the null hypothesis $\sigma = \sigma_0$ based on the statistic

$$z = \frac{\sqrt{2n}\,(s - \sigma_0)}{\sigma_0}$$

The null hypothesis is rejected for $z < -z_\alpha$ when the alternative hypothesis is $\sigma < \sigma_0$, for $z > z_\alpha$ when the alternative hypothesis is $\sigma > \sigma_0$, and for $z < -z_{\alpha/2}$ or $z > z_{\alpha/2}$ when the alternative hypothesis is $\sigma \neq \sigma_0$.

F.24 **F Test for the Null Hypothesis $\sigma_1 = \sigma_2$ (Normal Populations)**—A test of the null hypothesis $\sigma_1 = \sigma_2$ based on the statistic

$$F = \frac{s_2^2}{s_1^2}, \quad F = \frac{s_1^2}{s_2^2}, \quad \text{or} \quad F = \frac{s_2^2}{s_1^2} \text{ or } \frac{s_1^2}{s_2^2}$$

(whichever is larger)

depending on whether the alternative hypothesis is $\sigma_1 < \sigma_2$, $\sigma_1 > \sigma_2$, or $\sigma_1 \neq \sigma_2$. In the first case the null hypothesis is rejected for $F > F_{\alpha,n_2-1,n_1-1}$, in the second case it is rejected for $F > F_{\alpha,n_1-1,n_2-1}$, and in the third case it is rejected for $F > F_{\alpha,\nu_1,\nu_2}$, where ν_1 and ν_2 equal $n_2 - 1$ and $n_1 - 1$ when $s_2^2 > s_1^2$, and $n_1 - 1$ and $n_2 - 1$ when $s_1^2 > s_2^2$.

G. FORMULAS FOR THE ANALYSIS OF COUNT DATA

In the formulas of this section all confidence limits have the degree of confidence $1 - \alpha$, and all tests have the level of significance α. The symbol z_α denotes the value for which the area to its right under the standard normal curve is equal to α, and $\chi^2_{\alpha,\nu}$ is defined correspondingly for the chi-square distribution with ν degrees of freedom. A general requirement for the use of the various chi-square statistics is that none of the expected frequencies be less than 5; this can sometimes be avoided by combining several of the classes.

G.1 **Standard Error of a Proportion**—If x is a value of a random variable having the binomial distribution,

$$\sigma_{x/n} = \sqrt{\frac{\theta(1-\theta)}{n}}$$

G.2 **Standard Error of a Proportion (Finite Population)**—If x is a value of a random variable having the hypergeometric distribution,

$$\sigma_{x/n} = \sqrt{\frac{\theta(1-\theta)}{n} \cdot \frac{N-n}{N-1}}$$

where $N = a + b$ and $\theta = \dfrac{a}{a+b}$. This formula is used when sampling is *without replacement*; formula G.1 is used for sampling *with replacement* from a finite population.

G.3 **Standard Error of the Difference Between Two Proportions**—If x_1 and x_2 are values of independent random variables having binomial distributions with the respective parameters n_1 and θ_1, and n_2 and θ_2,

$$\sigma_{\frac{x_1}{n_1} - \frac{x_2}{n_2}} = \sqrt{\frac{\theta_1(1-\theta_1)}{n_1} + \frac{\theta_2(1-\theta_2)}{n_2}}$$

G.4 **Maximum Error of Estimate of Population Proportion (Large Sample)**—When a sample proportion x/n is used as an estimate of θ, the parameter of a binomial distribution, one can assert with the probability $1 - \alpha$ that the error $E = |x/n - \theta|$ will not exceed

(G.4a) $$E = z_{\alpha/2}\sqrt{\frac{\theta(1-\theta)}{n}}$$

or approximately

$$z_{\alpha/2}\sqrt{\frac{\frac{x}{n}\left(1-\frac{x}{n}\right)}{n}}$$

Correspondingly, the minimum sample size required to attain this precision is

(G.4b) $$n = \theta(1-\theta)\frac{z^2_{\alpha/2}}{E^2} \leq \frac{z^2_{\alpha/2}}{4F^2}$$

G.5 **Confidence Limits for a Proportion (Large Sample)**—If x is a value of a random variable having the binomial distribution and n is large, approximate cofidence limits for the parameter θ are given by

$$\frac{x}{n} \pm z_{\alpha/2}\sqrt{\frac{\frac{x}{n}\left(1-\frac{x}{n}\right)}{n}}$$

For small samples, confidence limits for θ may be obtained with the use of special tables; for example, those given in the Pearson-Hartley Tables listed on page 195.

G.6 **Confidence Limits for a Proportion (Finite Population)**—If x is a value of a random variable having the hypergeometric distribution and n is large, approximate confidence limits for the parameter $\theta = \frac{a}{a+b}$ are given by

$$\frac{x}{n} \pm z_{\alpha/2}\sqrt{\frac{\frac{x}{n}\left(1-\frac{x}{n}\right)}{n}\cdot\frac{N-n}{N-1}}$$

where $N = a + b$. These limits apply to sampling *without replacement* from a finite population, whereas those of G.5 apply to sampling *with replacement.*

G.7 **Confidence Limits for Difference Between Proportions (Large Samples)** – If x_1 and x_2 are values of independent random variables having binomial distributions with the respective parameters n_1 and θ_1, and n_2 and θ_2, where n_1 and n_2 are large, approximate confidence limits for $\theta_1 - \theta_2$ are given by

$$\frac{x_1}{n_1} - \frac{x_2}{n_2} \pm z_{\alpha/2}\sqrt{\frac{\frac{x_1}{n_1}\left(1 - \frac{x_1}{n_1}\right)}{n_1} + \frac{\frac{x_2}{n_2}\left(1 - \frac{x_2}{n_2}\right)}{n_2}}$$

G.8 **Large-Sample Test for a Proportion** – An approximate test of the null hypothesis $\theta = \theta_0$, where θ is the parameter of a binomial distribution; it is based on the statistic

(G.8a) $$z = \frac{\frac{x}{n} - \theta_0}{\sqrt{\frac{\theta_0(1 - \theta_0)}{n}}} = \frac{x - n\theta_0}{\sqrt{n\theta_0(1 - \theta_0)}}$$

The null hypothesis is rejected for $z < -z_\alpha$ when the alternative hypothesis is $\theta < \theta_0$, for $z > z_\alpha$ when the alternative hypothesis is $\theta > \theta_0$, and for $z < -z_{\alpha/2}$ or $z > z_{\alpha/2}$ when the alternative hypothesis is $\theta \neq \theta_0$. For small samples, exact tests of this null hypothesis are based directly on tables of binomial probabilities; for example, those in the *National Bureau of Standards Table* listed among the references on page 194. A *continuity correction* (which corrects for the approximation of a discrete distribution with a continuous distribution) consists of replacing x in the numerator of the formula for z with $x - 1/2$ or $x + 1/2$, whichever makes z numerically smallest. If the alternative hypothesis is one-sided and the probability of a Type II error at $\theta = \theta_1$ is to be β, the required sample size is given by

(G.8b) $$n = \frac{\left[z_\alpha\sqrt{\theta_0(1 - \theta_0)} + z_\beta\sqrt{\theta_1(1 - \theta_1)}\right]^2}{(\theta_1 - \theta_0)^2}$$

For the two-tail test, $z_{\alpha/2}$ must be substituted in this formula for z_α.

G.9 **Large-Sample Test for a Proportion (Finite Population)** – An approximate test of the null hypothesis $\theta = \theta_0$, where $\theta = \frac{a}{a+b}$ is the parameter of a hypergeometric distribution; it is based on the statistic

$$z = \frac{\frac{x}{n} - \theta_0}{\sqrt{\frac{\theta_0(1-\theta_0)}{n} \cdot \frac{N-n}{N-1}}}$$

$$= \frac{x - n\theta_0}{\sqrt{\frac{n\theta_0(1-\theta_0)(N-n)}{N-1}}}$$

The null hypothesis is rejected for $z < -z_\alpha$ when the alternative hypothesis is $\theta < \theta_0$, for $z > z_\alpha$ when the alternative hypothesis is $\theta > \theta_0$, and for $z < -z_{\alpha/2}$ or $z > z_{\alpha/2}$ when the alternative hypothesis is $\theta \neq \theta_0$. This test applies to sampling *without replacement* from a finite population, and it should be used whenever the sample constitutes at least 5 per cent of the population. The continuity correction of **G.8** may also be applied.

G.10 **Large-Sample Test for Difference Between Proportions** – An approximate test of the null hypothesis $\theta_1 - \theta_2 = \delta$, where n_1 and θ_1, and n_2 and θ_2, are the respective parameters of two binomial distributions; it is based on the statistic

$$z = \frac{\frac{x_1}{n_1} - \frac{x_2}{n_2} - \delta}{\sqrt{\frac{\frac{x_1}{n_1}\left(1 - \frac{x_1}{n_1}\right)}{n_1} + \frac{\frac{x_2}{n_2}\left(1 - \frac{x_2}{n_2}\right)}{n_2}}}$$

where x_1 and x_2 are observed values of the two independent random variables. The null hypothesis is rejected for $z < -z_\alpha$

when the alternative hypothesis is $\theta_1 - \theta_2 < \delta$, for $z > z_\alpha$ when the alternative hypothesis is $\theta_1 - \theta_2 > \delta$, and for $z < -z_{\alpha/2}$ or $z > z_{\alpha/2}$ when the alternative hypothesis is $\theta_1 - \theta_2 \neq \delta$. The continuity correction of **G.8** may also be applied to both x_1 and x_2.

G.11 **Binomial Index of Dispersion** – A chi-square statistic given by

$$\chi^2 = \frac{\sum_{i=1}^{k}(x_i - \bar{x})^2}{\bar{x}\left(1 - \frac{\bar{x}}{n}\right)}$$

where $x_1, x_2, \ldots,$ and x_k are the values of k independent random variables having binomial distributions with the parameters n_i and θ_i, $i = 1, 2, \ldots,$ and k; $\bar{x}$ is the mean of these x's. The null hypothesis $\theta_1 = \theta_2 = \ldots = \theta_k$ is rejected for $\chi^2 > \chi^2_{\alpha,k-1}$.

G.12 **Poisson Index of Dispersion** – A chi-square statistic given by

$$\chi^2 = \frac{\sum_{i=1}^{k}(x_i - \bar{x})^2}{\bar{x}}$$

where $x_1, x_2, \ldots,$ and x_k are values of k independent random variables having Poisson distributions with the parameters $\lambda_1, \lambda_2, \ldots,$ and λ_k; $\bar{x}$ is the mean of these x's. The null hypothesis $\lambda_1 = \lambda_2 = \ldots = \lambda_k$ is rejected for $\chi^2 > \chi^2_{\alpha,k-1}$.

G.13 **Chi-Square Statistic (Analysis of Contingency Table)** – A chi-square statistic given by

(G.13a)
$$\chi^2 = \sum_{i=1}^{a}\sum_{j=1}^{b}\frac{(f_{ij} - e_{ij})^2}{e_{ij}}$$

where f_{ij} is the number of items in the cell which belongs to the ith row and the jth column, for $i = 1, 2, \ldots, a$, and $j = 1, 2, \ldots, b$. The *expected cell frequencies* e_{ij} are calculated by means of the formula

(G.13b) $$e_{ij} = \frac{(f_{i.})(f_{.j})}{n}$$

where $f_{i.}$ is the total for the ith row, $f_{.j}$ is the total for the jth column, and n is the total for the entire table. The null hypothesis that the random variables represented by the two classifications are independent [or the null hypothesis that the columns (or rows) represent independent random samples from identical multinomial populations] is rejected for $\chi^2 > \chi^2_{\alpha, (a-1)(b-1)}$.

G.14 **Contingency Coefficient**—A measure of the strength of the dependence of two random variables represented by the classifications of a contingency table, given by

$$C = \sqrt{\frac{\chi^2}{\chi^2 + n}}$$

where χ^2 is calculated according to (G.13a).

G.15 **Chi-Square Statistic (Analysis of 2 × 2 Table)**—A measure of strength of the dependence between two dichotomized variables, given by

$$\chi^2 = \frac{(a + b + c + d)(ad - bc)^2}{(a + b)(c + d)(b + d)(a + c)}$$

where a, b, c, and d are the cell frequencies shown in the following 2 × 2 table:

a	b
c	d

This statistic is a special case of (G.13a).

G.16 **Coefficient of Association (Analysis of 2 × 2 Table)**—A measure of the strength of the dependence between two dichotomized variables, given by

$$Q = \frac{ad - bc}{ad + bc}$$

where a, b, c, and d are as defined in G.15.

G.17 **Phi Coefficient (Analysis of 2 × 2 Table)**—A measure of the strength of the dependence between two dichotomized variables, given by

(G.17a) $$\varphi = \frac{bc - ad}{\sqrt{(a + b)(c + d)(a + c)(b + d)}}$$

where a, b, c, and d are as defined in G.15. If χ^2 has already been calculated for a 2 × 2 table, φ may be obtained by means of the formula

(G.17b) $$\varphi = \sqrt{\chi^2/n}$$

G.18 **Tetrachoric Correlation Coefficient (Analysis of 2 × 2 Table)**—A measure of the strength of the dependence between two dichotomized variables having the bivariate normal distribution; it is based on the ratio

$$k = \frac{bc}{ad} \text{ or } \frac{ad}{bc} \quad \text{(whichever is larger)}$$

where a, b, c, and d are as defined in G.15. Since the calculation of the tetrachoric correlation coefficient r_t is relatively complicated, its value is usually determined for a given value of k with the use of a special table, such as the one in the second book by A. L. Edwards listed on page 193.

G.19 **Chi-Square Statistic (Goodness of Fit)**—A chi-square statistic given by

$$\chi^2 = \sum_{i=1}^{k} \frac{(f_i - e_i)^2}{e_i}$$

where f_1, f_2, . . . , and f_k are the class frequencies of an observed distribution, while e_1, e_2, . . . , and e_k are the corresponding frequencies *expected* for a given probability function (or a given probability density). The null hypothesis that this probability function (or probability density) provides the appropriate model for the data is rejected for $\chi^2 > \chi^2_{\alpha,\nu}$, where ν equals k *minus* the number of quantities (determined from the observed data) which are required for the calculation of the expected frequencies.

H. NONPARAMETRIC TESTS

This section contains descriptions and formulas of some of the most popular nonparametric tests; others are described briefly in Part I. All of the tests have the level of significance α. The symbol z_α denotes the value for which the area to its right under the standard normal curve is equal to α, and $\chi^2_{\alpha,\nu}$ is defined correspondingly for the chi-square distribution with ν degrees of freedom. For small samples, most of the tests of this section have to be based on special tables; for example, those in the handbook by D. B. Owen listed on page 194.

H.1 **One-Sample Sign Test** – A test of the null hypothesis that the mean of a symmetrical population is $\mu = \mu_0$. Each sample value is replaced by a plus sign if it exceeds μ_0, by a minus sign if it is less than μ_0, and the test is based on the statistic

$$z = \frac{2x - n}{\sqrt{n}}$$

where x is the number of plus signs and $n - x$ the number of minus signs. For large samples, the null hypothesis is rejected for $z < -z_\alpha$ when the alternative hypothesis is $\mu < \mu_0$, for $z > z_\alpha$ when the alternative hypothesis is $\mu > \mu_0$, and for $z < -z_{\alpha/2}$ or $z > z_{\alpha/2}$ when the alternative hypothesis is $\mu \neq \mu_0$. For small samples, the test is based directly on tables of binomial probabilities with $\theta = 1/2$.

H.2 **Two-Sample Sign Test** – A test of the null hypothesis that two independent random samples come from identical populations. The values of the two samples are *randomly* paired, each pair is replaced by a plus sign if the value from the first sample is greater than that from the second sample, by a minus sign if the value from the first sample is less than that from the second sample, and the procedure is the same as in the one-sample sign test **H.1**. When the sample sizes are not equal, some of the values of the larger sample will have to be discarded.

H.3 **Paired-Sample Sign Test** – This test is identical with the two-sample sign test **H.2**, except that the data are actually given as *matched pairs*.

H.4 **Median Test** – A test of the null hypothesis that k independent random samples come from identical populations. For each sample one determines how many values fall below the median of the combined data and how many fall above it; then, these frequencies are analyzed as a $2 \times k$ contingency table and the test is based on the chi-square statistic (**G.13a**). The null hypothesis is rejected for $\chi^2 > \chi^2_{\alpha, k-1}$.

H.5 **Mann-Whitney Test** – A test of the null hypothesis that two independent random samples of size n_1 and n_2 come from identical populations. The sample values are ranked jointly from 1 to $n_1 + n_2$, and the test is based on the statistic

(H.5a) $$U = n_1 n_2 + \frac{n_1(n_1 + 1)}{2} - R_1$$

where R_1 is the sum of the ranks assigned to the values of the first sample. The sampling distribution of this statistic has the mean

(H.5b) $$\mu_U = \frac{n_1 n_2}{2}$$

and the standard deviation

(H.5c) $$\sigma_U = \sqrt{\frac{n_1 n_2(n_1 + n_2 + 1)}{12}}$$

When n_1 and n_2 are both greater than 8, an approximate test can be based on the statistic

(H.5d) $$z = \frac{U - \mu_U}{\sigma_U}$$

in which case the null hypothesis is rejected for $z < -z_{\alpha/2}$ or $z > z_{\alpha/2}$. For small samples, the test is based on special tables.

H.6 **Kruskal-Wallis Test** – A test of the null hypothesis that k independent random samples of size $n_1, n_2, \ldots,$ and n_k, come from identical continuous populations. The sample values are ranked jointly from 1 to

$$\sum_{i=1}^{k} n_i = n,$$

and the test is based on the statistic

$$H = \frac{12}{n(n+1)} \sum_{i=1}^{k} \frac{R_i^2}{n_i} - 3(n+1)$$

where R_i is the sum of the ranks assigned to the values of the ith sample. When the sample sizes are all greater than 5, the null hypothesis is rejected for $H > \chi^2_{\alpha,k-1}$; for small samples, the test is based on special tables.

H.7 **Wilcoxon Test**—A test of the null hypothesis that the respective members of n matched pairs come from identical populations. The absolute values of the differences between the pairs are ranked from 1 to n, and the test is based on the statistic T, the sum of the ranks assigned to all the positive differences *or* all the negative differences, whichever are less frequent. The sampling distribution of this statistic has the mean

(H.7a) $$\mu_T = \frac{n(n+1)}{4}$$

and the standard deviation

(H.7b) $$\sigma_T = \sqrt{\frac{n(n+1)(2n+1)}{24}}$$

When n is greater than 25 the test is based on the statistic

(H.7c) $$z = \frac{T - \mu_T}{\sigma_T}$$

and the null hypothesis is rejected for $z < -z_{\alpha/2}$ or $z > z_{\alpha/2}$; for small samples, the test is based on special tables.

H.8 **Total Number of Runs** — A test of the randomness of a sequence of two kinds of symbols, based on the total number of runs, u. The sampling distribution of this statistic has the mean

(H.8a) $$\mu_u = \frac{2n_1n_2}{n_1 + n_2} + 1$$

where n_1 is the number of symbols of one kind and n_2 is the number of symbols of the other kind, and the standard deviation

(H.8b) $$\sigma_u = \sqrt{\frac{2n_1n_2(2n_1n_2 - n_1 - n_2)}{(n_1 + n_2)^2(n_1 + n_2 - 1)}}$$

When n_1 and n_2 are both greater than 10, the test is based on the statistic

(H.8c) $$z = \frac{u - \mu_u}{\sigma_u}$$

and the null hypothesis is rejected for $z < -z_{\alpha/2}$ or $z > z_{\alpha/2}$; for small samples, the test is based on special tables.

H.9 **Runs Above and Below the Median** — A test of the randomness of a sample based on the order in which the sample values were obtained. Each value is replaced by the letter a or the letter b depending on whether it is above or below the median of the sample; then, the run test **H.8** is applied to the resulting sequence of a's and b's.

H.10 **Wald-Wolfowitz Runs Test** — A test of the null hypothesis that two random samples come from identical populations. The data are arranged jointly according to size, each value is replaced by a 1 or a 2 depending on the sample to which it belongs, and the run test **H.8** is applied to the resulting sequence of 1s and 2s.

H.11 **Kolmogorov-Smirnov One-Sample Test** — A test of the null hypothesis that a given continuous distribution function with values $F(x)$ represents the population from which a random sample of size n was obtained. It is based on the maximum difference

$$D = \max_{x} | F(x) - F_n(x) |$$

where $F_n(x)$ denotes the proportion of the sample values that are less than or equal to x. The critical values of the test are obtained from special tables.

H.12 **Kolmogorov-Smirnov Two-Sample Test** — A test of the null hypothesis that two independent random samples of size n come from identical continuous populations. It is based on the maximum difference

$$D = \max_{x} | F_n(x) - G_n(x) |$$

where $F_n(x)$ denotes the proportion of the values in the first sample that are less than or equal to x, and $G_n(x)$ is defined correspondingly for the second sample. The critical values of the test are obtained from special tables.

I. CURVE FITTING, REGRESSION, AND CORRELATION

Formulas I.1 through I.30 apply to n pairs of observations (x_1, y_1), (x_2, y_2), . . . , and (x_n, y_n). Under the assumptions of normal regression analysis, the x's are constants and the y's are values of independent random variables having normal distributions with the respective means $\alpha + \beta x_i$ and the common variance σ^2. Under the assumptions of normal correlation analysis, the pairs (x_i, y_i) constitute a random sample from a bivariate normal population. The confidence intervals all have the degree of confidence $1 - \alpha$, the limits of prediction have the probability $1 - \alpha$, and all tests have the level of significance α. The symbol z_α denotes the value for which the area to its right under the standard normal curve is equal to α, and $t_{\alpha,\nu}$ is defined correspondingly for the t distribution with ν degrees of freedom.

I.1 Curve Fitting – Straight Line

(I.1a)
$$y = a + bx$$

For a line fit by the method of least squares, the values of a and b are obtained by solving the system of *normal equations*

(I.1b)
$$\Sigma y = na + b(\Sigma x)$$
$$\Sigma xy = a(\Sigma x) + b(\Sigma x^2)$$

Symbolically, the solutions of these equations are

(I.1c)
$$a = \frac{(\Sigma x^2)(\Sigma y) - (\Sigma x)(\Sigma xy)}{n(\Sigma x^2) - (\Sigma x)^2}$$

(I.1d)
$$b = \frac{n(\Sigma xy) - (\Sigma x)(\Sigma y)}{n(\Sigma x^2) - (\Sigma x)^2}$$

If b is calculated first, the following is an alternate formula for a

(I.1e)
$$a = \frac{\Sigma y - b(\Sigma x)}{n} = \bar{y} - b \cdot \bar{x}$$

I.2 Curve Fitting – Exponential Curve

(I.2a)
$$y = ab^x$$

or, in logarithmic form,

(I.2b) $\log y = \log a + x(\log b)$

Fitting an exponential curve by the method of least squares is equivalent to fitting a straight line I.1 to the points $(x_i, \log y_i)$.

I.3 Curve Fitting – Power Function

(I.3a) $y = ax^b$

or, in logarithmic form,

(I.3b) $\log y = \log a + b(\log x)$

Fitting a power function by the method of least squares is equivalent to fitting a straight line I.1 to the points $(\log x_i, \log y_i)$.

I.4 Curve Fitting – Polynomial Function

(I.4a) $y = b_0 + b_1 x + b_2 x^2 + \ldots + b_k x^k$

For a polynomial function fit by the method of least squares, the values of $b_0, b_1, b_2, \ldots,$ and b_k are obtained by solving the system of $k + 1$ *normal equations*

(I.4b)

$$\begin{aligned}
\Sigma\, y &= nb_0 + b_1(\Sigma\, x) + b_2(\Sigma\, x^2) + \ldots + b_k(\Sigma\, x^k) \\
\Sigma\, xy &= b_0(\Sigma\, x) + b_1(\Sigma\, x^2) + b_2(\Sigma\, x^3) + \ldots + b_k(\Sigma\, x^{k+1}) \\
&\cdot\ \cdot\ \cdot\ \cdot\ \cdot\ \cdot\ \cdot\ \cdot\ \cdot\ \cdot\ \cdot\ \cdot\ \cdot\ \cdot \\
\Sigma\, x^k y &= b_0(\Sigma\, x^k) + b_1(\Sigma\, x^{k+1}) + b_2(\Sigma\, x^{k+2}) + \ldots + b_k(\Sigma\, x^{2k})
\end{aligned}$$

I.5 Curve Fitting – Modified Exponential Curve

$$y = c + ab^x$$

One (of many) ways of fitting a modified exponential curve is to select three points whose coordinates are known, or can be estimated from the data, *with the x's equally spaced*; these values are substituted into the equation $y = c + ab^x$ and the resulting system of equations is solved for a, b, and c. If

$b < 1$, an alternate way is to determine the asymptotic value c (the value approached for large x) by inspection, and then apply the method of I.2 to the points $(x_i, y_i - c)$.

I.6 Curve Fitting – Logistic Curve

$$y = \frac{1}{c + ab^x}$$

Fitting a logistic curve is equivalent to fitting a modified exponential curve I.5 to the points $(x_i, 1/y_i)$.

I.7 Curve Fitting – Gompertz Curve

(I.7a) $$y = C \cdot A^{b^x}$$

or in logarithmic form

(I.7b) $$\log y = c + ab^x$$

where $c = \log C$ and $a = \log A$. Fitting a Gompertz curve is equivalent to fitting a modified exponential curve I.5 to the points $(x_i, \log y_i)$.

I.8 Regression Line – For the regression of Y on x,

$$\mu_{Y|x} = \alpha + \beta x$$

where $\mu_{Y|x}$ is the mean of the distribution of Y for a given x.

I.9 Standard Error of Estimate – Under the assumptions of normal regression analysis, an estimate of σ given by

(I.9a) $$s_e = \sqrt{\frac{\Sigma [y - (a + bx)]^2}{n - 2}}$$

where a and b are calculated according to the method of I.1. The following is a short-cut computing formula

(I.9b) $$s_e = \sqrt{\frac{\Sigma\, y^2 - a(\Sigma\, y) - b(\Sigma\, xy)}{n - 2}}$$

I.10 Confidence Limits for the Regression Coefficient α — Under the assumptions of normal regression analysis,

$$a \pm t_{\alpha/2,n-2} \cdot s_e \sqrt{\frac{1}{n} + \frac{n \cdot \bar{x}^2}{n(\Sigma x^2) - (\Sigma x)^2}}$$

where a is calculated by the method of I.1 and s_e is given by I.9.

I.11 Confidence Limits for the Regression Coefficient β — Under the assumptions of normal regression analysis,

$$b \pm \frac{t_{\alpha/2,n-2} \cdot s_e}{\sqrt{\dfrac{n(\Sigma x^2) - (\Sigma x)^2}{n}}}$$

where b is calculated by the method of I.1 and s_e is given by I.9.

I.12 Confidence Limits for $\alpha + \beta x_0$ — Under the assumptions of normal regression analysis,

$$(a + bx_0) \pm t_{\alpha/2,n-2} \cdot s_e \sqrt{\frac{1}{n} + \frac{n(x_0 - \bar{x})^2}{n(\Sigma x^2) - (\Sigma x)^2}}$$

where x_0 is given, a and b are obtained by the method of I.1, and s_e is given by I.9.

I.13 Limits of Prediction for a Value of Y at x_0 — Under the assumptions of normal regression analysis,

$$(a + bx_0) \pm t_{\alpha/2,n-2} \cdot s_e \sqrt{1 + \frac{1}{n} + \frac{n(x_0 - \bar{x})^2}{n(\Sigma x^2) - (\Sigma x)^2}}$$

where x_0 is given, a and b are obtained by the method of I.1, and s_e is given by I.9.

I.14 **t Test for the Regression Coefficient α** – Under the assumptions of normal regression analysis, a test of the null hypothesis $\alpha = \alpha_0$ based on the statistic

$$t = \frac{a - \alpha_0}{s_e \sqrt{\frac{1}{n} + \frac{n \cdot \bar{x}^2}{n(\Sigma x^2) - (\Sigma x)^2}}}$$

where a is obtained by the method of **I.1** and s_e is given by **I.9**. The null hypothesis is rejected for $t < -t_{\alpha,n-2}$ when the alternative hypothesis is $\alpha < \alpha_0$, for $t > t_{\alpha,n-2}$ when the alternative hypothesis is $\alpha > \alpha_0$, and for $t < -t_{\alpha/2,n-2}$ or $t > t_{\alpha/2,n-2}$ when the alternative hypothesis is $\alpha \neq \alpha_0$.

I.15 **t Test for the Regression Coefficient β** – Under the assumptions of normal regression analysis, a test of the null hypothesis $\beta = \beta_0$ based on the statistic

$$t = \frac{b - \beta_0}{s_e} \sqrt{\frac{n(\Sigma x^2) - (\Sigma x)^2}{n}}$$

where b is obtained by the method of **I.1** and s_e is given by **I.9**. The null hypothesis is rejected for $t < -t_{\alpha,n-2}$ when the alternative hypothesis is $\beta < \beta_0$, for $t > t_{\alpha,n-2}$ when the alternative hypothesis is $\beta > \beta_0$, and for $t < -t_{\alpha/2,n-2}$ or $t > t_{\alpha/2,n-2}$ when the alternative hypothesis is $\beta \neq \beta_0$.

I.16 **t Test for the Mean $\mu_{Y|x_0} = \alpha + \beta x_0$** – Under the assumptions of normal regression analysis, a test of the null hypothesis $\mu_{Y|x_0} = \mu_0$ for given x_0, based on the statistic

$$t = \frac{a + bx_0 - \mu_0}{s_e \sqrt{\frac{1}{n} + \frac{n(x_0 - \bar{x})^2}{n(\Sigma x^2) - (\Sigma x)^2}}}$$

where a and b are obtained by the method of **I.1** and s_e is given by **I.9**. The null hypothesis is rejected for $t < -t_{\alpha,n-2}$

when the alternative hypothesis is $\mu_{Y|x_0} < \mu_0$, for $t > t_{\alpha,n-2}$ when the alternative hypothesis is $\mu_{Y|x_0} > \mu_0$, and for $t < -t_{\alpha/2,n-2}$ or $t > t_{\alpha/2,n-2}$ when the alternative hypothesis is $\mu_{Y|x_0} \neq \mu_0$.

I.17 **Coefficient of Correlation** – Under the assumptions of normal correlation analysis, an estimate of ρ given by

(I.17a)
$$r = \frac{\Sigma (x - \bar{x})(y - \bar{y})}{\sqrt{\Sigma (x - \bar{x})^2}\sqrt{\Sigma (y - \bar{y})^2}}$$

or the equivalent computing formula

(I.17b)
$$r = \frac{n(\Sigma xy) - (\Sigma x)(\Sigma y)}{\sqrt{n(\Sigma x^2) - (\Sigma x)^2}\sqrt{n(\Sigma y^2) - (\Sigma y)^2}}$$

An alternate formula, often used to *define* r, is given by

(I.17c)
$$r = \pm \sqrt{1 - \frac{\Sigma [y - (a + bx)]^2}{\Sigma (y - \bar{y})^2}}$$

where a and b are obtained by the method of **I.1**, and the sign of r is taken to be the same as that of b.

I.18 **Coefficient of Correlation (Grouped Data)**

(I.18a)
$$r = \frac{n(\Sigma xyf) - (\Sigma xf_x)(\Sigma yf_y)}{\sqrt{n(\Sigma x^2 f_x) - (\Sigma xf_x)^2}\sqrt{n(\Sigma y^2 f_y) - (\Sigma yf_y)^2}}$$

where f_x and f_y denote the frequencies corresponding to the class marks x and y, and f denotes the frequency of the corresponding cell of the correlation table. With *coding*, the formula becomes

(I.18b)
$$r = \frac{n(\Sigma uvf) - (\Sigma uf_u)(\Sigma vf_v)}{\sqrt{n(\Sigma u^2 f_u) - (\Sigma uf_u)^2}\sqrt{n(\Sigma v^2 f_v) - (\Sigma vf_v)^2}}$$

where the u's and v's are coded class marks obtained by replacing the x's and y's, respectively, with the integers $\ldots, -3, -2, -1, 0, 1, 2, 3, \ldots$. The frequencies f_u and f_v are defined analogous to f_x and f_y.

I.19 Coefficient of Alienation

$$\sqrt{1 - r^2}$$

where r is the coefficient of correlation.

I.20 Coefficient of Determination

$$r^2$$

where r is the coefficient of correlation.

I.21 Coefficient of Nondetermination

$$1 - r^2$$

where r is the coefficient of correlation

I.22 *z*-Transformation

(I.22a) $$z = \frac{1}{2} \ln \frac{1 + r}{1 - r}$$

where ln denotes *natural logarithm*, that is, logarithm to the base e. Under the assumptions of normal correlation analysis, the sampling distribution of this statistic is approximately normal with the mean

(I.22b) $$\mu_z = \frac{1}{2} \ln \frac{1 + \rho}{1 - \rho}$$

and the standard deviation

(I.22c) $$\sigma_z = \frac{1}{\sqrt{n - 3}}$$

I.23 **Standard Error of the Correlation Coefficient**—Under the assumptions of normal correlation analysis, a large-sample approximation is given by

$$\sigma_r = \frac{1 - \rho^2}{\sqrt{n - 1}}$$

I.24 **Confidence Limits for ρ (Large Samples)**—Under the assumptions of normal correlation analysis,

$$\frac{1 + r - (1 - r)e^{\pm 2z_{\alpha/2}/\sqrt{n-3}}}{1 + r + (1 - r)e^{\pm 2z_{\alpha/2}/\sqrt{n-3}}}$$

where r is the coefficient of correlation. For small samples, confidence limits for ρ may be obtained from special tables; for example, those in the Pearson-Hartley Tables listed on page 195.

I.25 **t Test for the Hypothesis $\rho = 0$**—Under the assumptions of normal correlation analysis, a test of the null hypothesis $\rho = 0$ based on the statistic

$$t = \frac{r\sqrt{n - 2}}{\sqrt{1 - r^2}}$$

where r is the coefficient of correlation. The null hypothesis is rejected for $t < -t_{\alpha,n-2}$ when the alternative hypothesis is $\rho < 0$, for $t > t_{\alpha,n-2}$ when the alternative hypothesis is $\rho > 0$, and for $t < -t_{\alpha/2,n-2}$ or $t > t_{\alpha/2,n-2}$ when the alternative hypothesis is $\rho \neq 0$.

I.26 **Large-Sample Test for ρ**—Under the assumptions of normal correlation analysis, a test of the null hypothesis $\rho = \rho_0$ based on the statistic

$$z = \frac{\sqrt{n - 3}}{2} \ln \frac{(1 + r)(1 - \rho_0)}{(1 - r)(1 + \rho_0)}$$

where r is the coefficient of correlation and ln denotes natural logarithm, that is, logarithm to the base e. The null hypothesis is rejected for $z < -z_\alpha$ when the alternative hypothesis $\rho < \rho_0$, for $z > z_\alpha$ when the alternative hypothesis is $\rho > \rho_0$, and for $z < -z_{\alpha/2}$ or $z > z_{\alpha/2}$ when the alternative hypothesis is $\rho \neq \rho_0$.

I.27 **Point Biserial Coefficient of Correlation**—A special form of the correlation coefficient, used when the x's are 0s and 1s corresponding to the two categories of a *dichotomous* variable; it is given by

$$r_{pb} = \frac{n(\Sigma' y) - n_1(\Sigma y)}{\sqrt{n_0 n_1 [n(\Sigma y^2) - (\Sigma y)^2]}}$$

where n_0 is the number of 0s, n_1 is the number of 1s, and $\Sigma' y$ is the sum of the y's for which $x = 1$.

I.28 **Biserial Coefficient of Correlation**—A special form of the correlation coefficient used when the x's are 0s and 1s corresponding to the two categories of an *artificially-dichotomized* continuous variable; it is given by

$$r_b = \frac{n(\Sigma' y) - n_1(\Sigma y)}{n y_p \sqrt{n(\Sigma y^2) - (\Sigma y)^2}}$$

where n_1 is the number of 1s, $\Sigma' y$ is the sum of the y's for which $x = 1$, and y_p is the ordinate of the standard normal curve at the point for which the area under the curve to the left (or to the right) equals $p = n_1/n$. Tables of the ordinates of the standard normal distribution may be found in most statistical tables.

I.29 **Spearman's Rank Correlation Coefficient**—A measure of correlation based on the rankings of the x's and the y's within their respective samples; it is given by

(I.29a) $$r_S = 1 - \frac{6(\Sigma d^2)}{n(n^2 - 1)}$$

where the d's are the differences between the ranks of the respective pairs of observations. In case of ties, tied observations are assigned the mean of the ranks they jointly occupy. When the x's and the y's are values of independent continuous random variables, the sampling distribution of r_S has the mean 0 and the standard deviation

(I.29b) $$\sigma_{r_S} = \frac{1}{\sqrt{n-1}}$$

A significance test for r_S may thus be based on the statistic

(I.29c) $$z = r_S\sqrt{n-1}$$

which, for *large samples*, has approximately the standard normal distribution. For small samples, a corresponding test may be based on the statistic

(I.29d) $$t = \frac{r_S\sqrt{n-2}}{\sqrt{1-r_S^2}}$$

which has approximately the t distribution with $n - 2$ degrees of freedom.

I.30 **Kendall's Tau** – A measure of correlation based on the rankings of the x's and the y's within their respective samples; if a pair of points (x_j, y_j) and (x_k, y_k) represents an *inversion* when x_j is greater than x_k while y_j is less than y_k (or vice versa), it is given by

(I.30a) $$\tau = 1 - \frac{4Q}{n(n-1)} = \frac{4P}{n(n-1)} - 1 = \frac{2S}{n(n-1)}$$

where Q is the *total number of inversions* among all possible pairs of data points, P is the number of pairs of data points which are not inversions, and $S = P - Q$ is the total score obtained by assigning +1 to each pair which represents an inversion and −1 to each pair which does not. In case of ties, tied observations are assigned the mean of the ranks they jointly occupy, and the formula for τ becomes

(I.30b)
$$\tau = \frac{S}{\sqrt{\frac{1}{2}n(n-1) - T}\,\sqrt{\frac{1}{2}n(n-1) - U}}$$

where $T = \frac{1}{2}[\Sigma\, t(t-1)]$, with t being the number of tied observations in each set of ties among the x's, and $U = \frac{1}{2}[\Sigma\, t(t-1)]$, with t being the number of tied observations in each set of ties among the y's. Under the assumption that the x's and the y's are values of independent continuous random variables, the sampling distribution of τ has the mean 0 and the standard deviation

(I.30c)
$$\sigma = \sqrt{\frac{2(2n+5)}{9n(n-1)}}$$

A significance test for τ may thus be based on the statistic

(I.30d)
$$z = \tau\sqrt{\frac{9n(n-1)}{2(2n+5)}}$$

which, for *large samples*, has approximately the standard normal distribution. For small samples, such a test is based on special tables.

I.31 Coefficient of Concordance — A measure of the agreement among k rankings of n individuals or items, given by

(I.31a)
$$W = \frac{12(\Sigma\, R_i^2)}{k^2 n(n^2-1)} - \frac{3(n+1)}{n-1}$$

where R_i is the sum of the ranks assigned to the ith item. For $n > 7$, a significance test of the null hypothesis that the rankings are independent may be based on the statistic

(I.31b)
$$\chi^2 = k(n-1)W$$

which has approximately the chi-square distribution with $n - 1$ degrees of freedom. For smaller samples, such a test is based on special tables.

I.32 Curve Fitting – Linear Equation in $k + 1$ Unknowns

(I.32a) $$y = b_0 + b_1x_1 + b_2x_2 + \ldots + b_kx_k$$

For n data points $(x_1, x_2, \ldots, x_k, y)$, least squares values of the coefficients b_0, b_1, b_2, ..., and b_k are obtained by solving the system of $k + 1$ *normal equations*

$$\Sigma\, y = nb_0 + b_1(\Sigma\, x_1) + b_2(\Sigma\, x_2) + \ldots + b_k(\Sigma\, x_k)$$

$$\Sigma\, x_1y = b_0(\Sigma\, x_1) + b_1(\Sigma\, x_1^2) + b_2(\Sigma\, x_1x_2) + \ldots + b_k(\Sigma\, x_1x_k)$$

$$\Sigma\, x_2y = b_0(\Sigma\, x_2) + b_1(\Sigma\, x_1x_2) + b_2(\Sigma\, x_2^2) + \ldots + b_k(\Sigma\, x_2x_k)$$

$$\cdot \quad \cdot \quad \cdot \quad \cdot \quad \cdot \quad \cdot \quad \cdot \quad \cdot \quad \cdot \quad \cdot \quad \cdot \quad \cdot \quad \cdot \quad \cdot \quad \cdot$$

$$\Sigma\, x_ky = b_0(\Sigma\, x_k) + b_1(\Sigma\, x_1x_k) + b_2(\Sigma\, x_2x_k) + \ldots + b_k(\Sigma\, x_k^2)$$

I.33 Partial Correlation Coefficient – For n data points $(x_1, x_2, \ldots x_k)$, the partial correlation coefficient for variables x_1 and x_2 is given by

(I.33a) $$r_{12.34\ldots k} = -\frac{R_{12}}{\sqrt{R_{11}R_{22}}}$$

where R_{12}, R_{11}, and R_{22} are *cofactors* of the determinant R of the pairwise correlation coefficients r_{ij} of variables x_i and x_j, namely,

(I.33b) $$R = \begin{vmatrix} 1 & r_{12} & r_{13} & \ldots & r_{1k} \\ r_{21} & 1 & r_{23} & \ldots & r_{2k} \\ \cdot & \cdot & \cdot & \cdot & \cdot \\ \cdot & \cdot & \cdot & \cdot & \cdot \\ r_{k1} & r_{k2} & r_{k3} & \ldots & 1 \end{vmatrix}$$

The *cofactor* R_{ij} is given by $(-1)^{i+j}$ times the value of the determinant obtained by deleting from R the ith row and the jth column. Corresponding partial correlation coefficients for any other pair of variables can be obtained by appropriately changing the subscripts throughout formula (**I.33a**). For $k = 3$, the formula for the partial correlation coefficient (**I.33a**) becomes

(I.33c) $$r_{12.3} = \frac{r_{12} - r_{13}r_{23}}{\sqrt{(1 - r_{13}^2)(1 - r_{23}^2)}}$$

I.34 **Multiple Correlation Coefficient**—For n data points $(x_1, x_2, \ldots, x_k)$, the multiple correlation coefficient measures the strength of the relationship between any one of the variables and a linear combination of the other $k - 1$; for x_1 it is given by

$$r_{1.23\ldots k} = \sqrt{1 - \frac{R}{R_{11}}}$$

where R and R_{11} are as defined in **I.33**. The other multiple correlation coefficients, $r_{2.13\ldots k}$, $r_{3.124\ldots k}$, $\ldots$, and $r_{k.12\ldots(k-1)}$, are obtained by appropriately changing the subscripts in the above formula.

J. ANALYSIS OF VARIANCE

Included in this section are only some of the most basic of analysis of variance techniques. The various F tests are performed at the level of significance α, and F_{α, ν_1, ν_2} denotes the value for which the area to its right under the F distribution with ν_1 and ν_2 degrees of freedom is equal to α. Estimates of the parameters are indicated by placing carets (^) on their respective symbols.

J.1 **One-Way Analysis of Variance** – The model equation for this kind of analysis is

(J.1a) $$y_{ij} = \mu + \alpha_i + e_{ij} \quad \text{for } i = 1, 2, \ldots, k; \; j = 1, 2, \ldots, n$$

Here y_{ij} is the jth observation for the ith treatment, μ is the *grand mean*, the *treatment effects* α_i are subject to the restriction

$$\sum_{i=1}^{k} \alpha_i = 0,$$

and the e_{ij} are values of independent random variables having identical normal distributions with zero means and the variance σ^2. To test the null hypothesis $\alpha_1 = \alpha_2 = \ldots = \alpha_k$ against the alternative that these treatment effects are *not all equal*, the following calculations are required:

Correction term

(J.1b) $$C = T_{..}^2/nk \quad \text{where } T_{..} = \sum_{i=1}^{k} \sum_{j=1}^{n} y_{ij}$$

Total sum of squares

(J.1c) $$SST = \sum_{i=1}^{k} \sum_{j=1}^{n} y_{ij}^2 - C$$

Treatment sum of squares

(J.1d) $$SS(Tr) = \sum_{i=1}^{k} T_{i.}^2/n - C \quad \text{where } T_{i.} = \sum_{j=1}^{n} y_{ij}$$

Error sum of squares

(J.1e) $$SSE = SST - SS(Tr)$$

Analysis of variance table

(J.1f)

Source of variation	Degrees of freedom	Sum of squares	Mean square	F
Treatments	$k-1$	$SS(Tr)$	$MS(Tr) = SS(Tr)/(k-1)$	$\frac{MS(Tr)}{MSE}$
Error	$k(n-1)$	SSE	$MSE = SSE/k(n-1)$	
Total	$nk-1$	SST		

The null hypothesis is rejected for $F > F_{\alpha, k-1, k(n-1)}$. Least squares estimates of the parameters are given by

(J.1g) $$\hat{\mu} = T_{..}/kn$$

(J.1h) $$\hat{\alpha}_i = T_{i.}/n - T_{..}/kn\,.$$

J.2 **One-Way Analysis of Variance (Sample Sizes Unequal)** – When there are n_i observations corresponding to the ith treatment, the analysis is like **J.1**, except that

$$N = \sum_{i=1}^{k} n_i$$

must be substituted for nk throughout, and the computing formulas (**J.1b**), (**J.1c**), and (**J.1d**) become

(J.2a) $$C = T_{..}^2/N \quad \text{where } T_{..} = \sum_{i=1}^{k} \sum_{j=1}^{n_i} y_{ij}$$

(J.2b) $$SST = \sum_{i=1}^{k} \sum_{j=1}^{n_i} y_{ij}^2 - C$$

(J.2c) $$SS(Tr) = \sum_{i=1}^{k} T_{i.}^2/n_i - C \quad \text{where } T_{i.} = \sum_{j=1}^{n_i} y_{ij}$$

J.3 **Two-Way Analysis of Variance** – The model equation for this kind of analysis is

(J.3a) $$y_{ij} = \mu + \alpha_i + \beta_j + e_{ij} \quad \text{for } i = 1, 2, \ldots, k;\ j = 1, 2, \ldots, n$$

Here y_{ij} is the value obtained for the ith treatment in the jth block, μ is the *grand mean*, the *treatment effects* α_i are subject to the restriction

$$\sum_{i=1}^{k} \alpha_i = 0,$$

the *block effects* β_j are subject to the restriction

$$\sum_{j=1}^{k} \beta_j = 0,$$

and the e_{ij} are values of independent random variables having identical normal distributions with zero means and the variance σ^2. To test the null hypothesis $\alpha_1 = \alpha_2 = \ldots = \alpha_k$ against the alternative that these treatment effects are *not all equal*, and the null hypothesis $\beta_1 = \beta_2 = \ldots = \beta_n$ against the alternative that these block effects are *not all equal*, the following calculations are required:

Correction term

(J.3b) $$C = T_{..}^2/nk \quad \text{where } T_{..} = \sum_{i=1}^{k} \sum_{j=1}^{n} y_{ij}$$

Total sum of squares

(J.3c) $$SST = \sum_{i=1}^{k} \sum_{j=1}^{n} y_{ij}^2 - C$$

Treatment sum of squares

(J.3d) $$SS(Tr) = \sum_{i=1}^{k} T_{i.}^2/n - C \quad \text{where } T_{i.} = \sum_{j=1}^{n} y_{ij}$$

Block sum of squares

(J.3e) $$SS(Bl) = \sum_{j=1}^{n} T^2_{.j}/k - C \quad \text{where } T_{.j} = \sum_{i=1}^{k} y_{ij}$$

Error sum of squares

(J.3f) $$SSE = SST - SS(Tr) - SS(Bl)$$

Analysis of variance table

(J.3g)

Source of variation	Degrees of freedom	Sum of squares	Mean square	F
Treatments	$k-1$	$SS(Tr)$	$MS(Tr) = SS(Tr)/(k-1)$	$F_{Tr} = \frac{MS(Tr)}{MSE}$
Blocks	$n-1$	$SS(Bl)$	$MS(Bl) = SS(Bl)/(n-1)$	$F_{Bl} = \frac{MS(Bl)}{MSE}$
Error	$(k-1)(n-1)$	SSE	$MSE = \frac{SSE}{(k-1)(n-1)}$	
Total	$kn-1$	SST		

The null hypothesis for treatment effects is rejected for $F > F_{\alpha,k-1,(k-1)(n-1)}$, and the null hypothesis for block effects is rejected for $F > F_{\alpha,n-1,(k-1)(n-1)}$. Least squares estimates of the parameters are given by

(J.3h) $$\hat{\mu} = T_{..}/nk$$

(J.3i) $$\hat{\alpha}_i = T_{i.}/n - T_{..}/nk$$

(J.3j) $$\hat{\beta}_j = T_{.j}/k - T_{..}/nk$$

K. INDEX NUMBER FORMULAS

In the formulas of this section, p_0 denotes the price of a commodity in the base year (or period) while q_0 denotes the corresponding quantity (produced, sold, consumed, etc.); p_n denotes the price of the commodity in the given year (or period) while q_n denotes the corresponding quantity; and q_a denotes the corresponding quantity in some fixed year (or period) other than the base year or the given year. Except for K.12, the index number formulas all represent price indices, but they can easily be converted into quantity indices by substituting p's for q's and vice versa.

K.1 **Price Relative** – For a given commodity, the ratio p_n/p_0.

K.2 **Quantity Relative** – For a given commodity, the ratio q_n/q_0.

K.3 **Simple Aggregative Index**

$$I = \frac{\Sigma\, p_n}{\Sigma\, p_0} \cdot 100$$

K.4 **Mean of Price Relatives** – For k commodities,

$$I = \frac{\Sigma \dfrac{p_n}{p_0}}{k} \cdot 100 = \frac{\Sigma \dfrac{p_n}{p_0} \cdot 100}{k}$$

K.5 **Geometric Mean of Price Relatives** – For k commodities,

(K.5a) $$I = \sqrt[k]{\pi \frac{p_n}{p_0}} \cdot 100$$

where $\pi \dfrac{p_n}{p_0}$ denotes the product of the price relatives of the k commodities; in logarithmic form, the formula becomes

(K.5b) $$\log \frac{I}{100} = \frac{\Sigma \log p_n/p_0}{k}$$

K.6 Weighted Mean of Price Relatives

$$I = \frac{\Sigma \frac{p_n}{p_o} \cdot w}{\Sigma w} \cdot 100$$

where the w's are the weights assigned to the price relatives of the individual commodities.

K.7 Weighted Aggregative Index

$$I = \frac{\Sigma p_n \cdot w}{\Sigma p_o \cdot w} \cdot 100$$

where the w's are the weights assigned to the prices of the individual commodities.

K.8 Fixed-Weight Aggregative Index

$$I = \frac{\Sigma p_n q_a}{\Sigma p_o q_a} \cdot 100$$

K.9 Laspeyres' Index

$$I = \frac{\Sigma p_n q_o}{\Sigma p_o q_o} \cdot 100$$

K.10 Paasche's Index

$$I = \frac{\Sigma p_n q_n}{\Sigma p_o q_n} \cdot 100$$

K.11 Fisher's Ideal Index

$$I = \sqrt{\frac{\Sigma p_n q_n}{\Sigma p_o q_n} \cdot \frac{\Sigma p_n q_o}{\Sigma p_o q_o}} \cdot 100$$

K.12 Value Index

$$I = \frac{\Sigma p_n q_n}{\Sigma p_o q_o} \cdot 100$$

L. FORMULAS FOR CONTROL CHARTS

The notation used in connection with control charts differs somewhat from that used in other areas of statistics. The mean and the standard deviation of a population are denoted $\bar{x}'$ and σ', the parameter θ of the binomial distribution is denoted p', and the parameter λ of the Poisson distribution is denoted c'. Also, the standard deviation of a sample is denoted σ and it is defined as $\sqrt{\Sigma (x - \bar{x})^2/n}$. The formulas given here are all for three-sigma control limits, and if the sampling distribution of the statistic on which such a control chart is based is normal, the probability of obtaining a value outside the control limits by chance is 0.0027. Values of the constants A, A_1, A_2, c_2, B_1, B_2, B_3, B_4, d_2, D_1, D_2, D_3, and D_4 are given in the table on page 192 for samples of size $n = 2, 3, \ldots,$ and 15.

L.1 **Analysis of Past Data** – If k random samples of size n have the means $\bar{x}_i$, the standard deviations σ_i, and the ranges R_i, for $i = 1, 2, \ldots,$ and k, then

(L.1a)
$$\bar{\bar{x}} = \frac{\sum_{i=1}^{k} \bar{x}_i}{k}$$

(L.1b)
$$\bar{\sigma} = \frac{\sum_{i=1}^{k} \sigma_i}{k}$$

(L.1c)
$$\bar{R} = \frac{\sum_{i=1}^{k} R_i}{k}$$

If $x_1, x_2, \ldots,$ and x_k are the number of defectives in k samples of size n from a binomial population,

(L.1d)
$$\bar{p} = \frac{\sum_{i=1}^{k} x_i}{kn}$$

If c_1, c_2, . . . , and c_k are the number of defects observed in k samples from a Poisson population,

(L.1e) $$\bar{c} = \frac{\sum_{i=1}^{k} c_i}{k}$$

L.2 **$\bar{x}$ Chart ($\bar{x}'$ and σ' Given)**

Central Line: $\bar{x}'$

Control Limits: $\bar{x}' \pm A\sigma'$

L.3 **$\bar{x}$ Chart ($\bar{\bar{x}}$ and $\bar{\sigma}$ Determined from Past Data)**

Central Line: $\bar{\bar{x}}$

Control Limits: $\bar{\bar{x}} \pm A_1\bar{\sigma}$

L.4 **$\bar{x}$ Chart ($\bar{\bar{x}}$ and $\bar{R}$ Determined from Past Data)**

Central Line: $\bar{\bar{x}}$

Control Limits: $\bar{\bar{x}} \pm A_2\bar{R}$

L.5 **σ Chart (σ' Given)**

Central Line: $c_2\sigma'$

Control Limits: $B_1\sigma'$ and $B_2\sigma'$

L.6 **σ Chart ($\bar{\sigma}$ Determined from Past Data)**

Central Line: $\bar{\sigma}$

Control Limits: $B_3\bar{\sigma}$ and $B_4\bar{\sigma}$

L.7 **R Chart (σ' Given)**

Central Line: $d_2\sigma'$

Control Limits: $D_1\sigma'$ and $D_2\sigma'$

L.8 R Chart ($\bar{R}$ Determined from Past Data)

Central Line: $\bar{R}$

Control Limits: $D_3\bar{R}$ and $D_4\bar{R}$

L.9 Number of Defectives Chart (p' Given)

Central Line: np'

Control Limits: $np' \pm 3\sqrt{np'(1-p')}$

L.10 Number of Defectives Chart ($\bar{p}$ Determined from Past Data)

Central Line: $n\bar{p}$

Control Limits: $n\bar{p} \pm 3\sqrt{n\bar{p}(1-\bar{p})}$

L.11 p Chart (p' Given)

Central Line: p'

Control Limits: $p' \pm 3\sqrt{\dfrac{p'(1-p')}{n}}$

where $p = x/n$ is the proportion of defectives in a sample of size n.

L.12 p Chart ($\bar{p}$ Determined from Past Data)

Central Line: $\bar{p}$

Control Limits: $\bar{p} \pm 3\sqrt{\dfrac{\bar{p}(1-\bar{p})}{n}}$

where $p = x/n$ is the proportion of defectives in a sample of size n.

L.13 c Chart (c' Given)

Central Line: c'

Control Limits: $c' \pm 3\sqrt{c'}$

L.14 c Chart ($\bar{c}$ Determined from Past Data)

Central Line: $\bar{c}$

Control Limits: $\bar{c} \pm 3\sqrt{\bar{c}}$

TABLE OF CONTROL CHART CONSTANTS

CONTROL CHART CONSTANTS*

Number of observations in sample n	Chart for averages			Chart for standard deviations					Chart for ranges				
	Factors for control limits			Factor for central line	Factors for control limits				Factor for central line	Factors for control limits			
	A	A_1	A_2	c_2	B_1	B_2	B_3	B_4	d_2	D_1	D_2	D_3	D_4
2	2.121	3.760	1.880	0.5642	0	1.843	0	3.267	1.128	0	3.686	0	3.267
3	1.732	2.394	1.023	0.7236	0	1.858	0	2.568	1.693	0	4.358	0	2.575
4	1.500	1.880	0.729	0.7979	0	1.808	0	2.266	2.059	0	4.698	0	2.282
5	1.342	1.596	0.577	0.8407	0	1.756	0	2.089	2.326	0	4.918	0	2.115
6	1.225	1.410	0.483	0.8686	0.026	1.711	0.030	1.970	2.534	0	5.078	0	2.004
7	1.134	1.277	0.419	0.8882	0.105	1.672	0.118	1.882	2.704	0.205	5.203	0.076	1.924
8	1.061	1.175	0.373	0.9027	0.167	1.638	0.185	1.815	2.847	0.387	5.307	0.136	1.864
9	1.000	1.094	0.337	0.9139	0.219	1.609	0.239	1.761	2.970	0.546	5.394	0.184	1.816
10	0.949	1.028	0.308	0.9227	0.262	1.584	0.284	1.716	3.078	0.687	5.469	0.223	1.777
11	0.905	0.973	0.285	0.9300	0.299	1.561	0.321	1.679	3.173	0.812	5.534	0.256	1.744
12	0.866	0.925	0.266	0.9359	0.331	1.541	0.354	1.646	3.258	0.924	5.592	0.284	1.716
13	0.832	0.884	0.249	0.9410	0.359	1.523	0.382	1.618	3.336	1.026	5.646	0.308	1.692
14	0.802	0.848	0.235	0.9453	0.384	1.507	0.406	1.594	3.407	1.121	5.693	0.329	1.671
15	0.775	0.816	0.223	0.9490	0.406	1.492	0.428	1.572	3.472	1.207	5.737	0.348	1.652

*Reproduced, by permission, from the *ASTM Manual on Quality Control of Materials*, American Society for Testing and Materials, Philadelphia, Pa., 1951.

REFERENCES

Ackoff, R. L., (ed.): *Progress in Operations Research*, vol. I, New York: John Wiley and Sons, Inc., 1961.

Anderson, R. L., and T. A. Bancroft: *Statistical Theory in Research*, New York: McGraw-Hill Book Company, 1952.

Bartlett, M. S.: *An Introduction to Stochastic Processes*, Cambridge: Cambridge University Press, 1955.

Brown, R. G.: *Statistical Forecasting for Inventory Control*, New York: McGraw-Hill Book Company, 1959.

Burger, E.: *Introduction to the Theory of Games*, Englewood Cliffs, N.J.: Prentice-Hall, Inc., 1963.

Cattell, R. B.: *Factor Analysis*, New York: Harper and Row, 1952.

Cochran, W. G.: *Sampling Techniques*, 2d ed., New York: John Wiley and Sons, Inc., 1963.

Cochran, W. G., and G. M. Cox: *Experimental Designs*, 2d ed., New York: John Wiley and Sons, Inc., 1957.

David, H. A.: *The Method of Paired Comparisons*, New York: Hafner Publishing Company, Inc., 1963.

Dixon, W. J., and F. J. Massey: *Introduction to Statistical Analysis*, 2d ed., New York: McGraw-Hill Book Company, 1957.

Dodge, H. F., and H. G. Romig: *Sampling Inspection Tables*, New York: John Wiley and Sons, Inc., 1944.

Edwards, A. L.: *Statistical Analysis for Students in Psychology and Education*, New York: Holt, Rinehart and Winston, Inc., 1946.

Edwards, A. L.: *Statistical Methods for the Behavioral Sciences*, New York: Holt, Rinehart and Winston, Inc., 1954.

Emmens, C. W.: *Principles of Biological Assay*, London: Chapman and Hall, Ltd., 1948.

REFERENCES

Federer, W.T.: *Experimental Design, Theory and Application*, New York: The Macmillan Company, 1955.

Feller, W.: *An Introduction to Probability Theory and Its Applications*, vol. I, 2d ed., New York: John Wiley and Sons, Inc., 1957.

Finney, D. J.: *Statistical Methods in Biological Assay*, 2d ed., New York: Hafner Publishing Company, Inc., 1964.

Fisher, R. A.: *Statistical Methods and Scientific Inference*, New York: Hafner Publishing Company, Inc., 1956.

Fisher, R. A., and F. Yates: *Statistical Tables for Biological, Agricultural, and Medical Research*, 6th ed., New York: Hafner Publishing Company, Inc., 1963.

Freund, J. E.: *Mathematical Statistics*, Englewood Cliffs, N.J.: Prentice-Hall, Inc., 1962.

Freund, J. E., and F.J. Williams: *Modern Business Statistics*, Englewood Cliffs, N.J.: Prentice-Hall, Inc., 1958.

Garvin, W. W.: *Introduction to Linear Programming*, New York: McGraw-Hill Book Company, 1960.

Hoel, P. G.: *Introduction to Mathematical Statistics*, 3d ed., New York: John Wiley and Sons, Inc., 1962.

Keeping, E. S.: *Introduction to Statistical Inference*, Princeton, N.J.: D. Van Nostrand Company, Inc., 1962.

Kendall, M. G., and A. Stuart: *The Advanced Theory of Statistics*, vols. 1 and 2, London: Charles Griffin and Company, Ltd., 1958 and 1961.

Lindley, D. V.: *Introduction to Probability and Statistics from a Bayesian Viewpoint*, Parts 1 and 2, Cambridge: Cambridge University Press, 1965.

McKinsey, J. C. C.: *Introduction to the Theory of Games*, New York: McGraw-Hill Book Company, 1952.

Miller, I., and J. E. Freund: *Probability and Statistics for Engineers*, Englewood Cliffs, N.J.: Prentice-Hall, Inc., 1965.

Molina, E. C.: *Poisson's Exponential Binomial Limit*, Princeton, N.J.: D. Van Nostrand Company, Inc., 1947.

Mood, A. M., and F. A. Graybill: *Introduction to the Theory of Statistics*, 2d ed., New York: McGraw-Hill Book Company, 1963.

National Bureau of Standards: *Tables of the Binomial Distribution*, 1950.

Owen, D. B.: *Handbook of Statistical Tables*, Reading, Mass.: Addison-Wesley Publishing Company, Inc., 1962.

Parzen, E.: *Stochastic Processes*, San Francisco: Holden-Day, Inc., 1962.

Pearson, E. S., and H. O. Hartley: *Biometrika Tables for Statisticians*, Cambridge: Cambridge University Press, 1954.

RAND Corporation: *A Million Random Digits with 100,000 Normal Deviates*, New York: The Free Press, 1955.

Romig, H. G.: *50-100 Binomial Tables*, New York: John Wiley and Sons, Inc., 1953.

Sasieni, M., A. Yaspan, and L. Friedman: *Operations Research – Methods and Problems*, New York: John Wiley and Sons, Inc., 1959.

Savage, L. J.: *The Foundations of Statistical Inference*, New York: John Wiley and Sons, Inc., 1962.

Schlaifer, R.: *Probability and Statistics for Business Decisions*, New York: McGraw-Hill Book Company, 1959.

Siegel, S.: *Nonparametric Statistics for the Behavioral Sciences*, New York: McGraw-Hill Book Company, 1956.

Snedecor, G. W.: *Statistical Methods*, 5th ed., Ames, Iowa: Iowa State University Press, 1956.

Stead, W. H., C. L. Shartle et al.: *Occupational Counseling Techniques*, New York: American Book Company, 1940.

Wald, A.: *On the Principles of Statistical Inference*, South Bend, Ind.: Notre Dame University, 1942.

Wilks, S. S.: *Mathematical Statistics*, New York: John Wiley and Sons, Inc., 1962.

Wine, R. L.: *Statistics for Scientists and Engineers*, Englewood Cliffs, N.J.: Prentice-Hall, Inc., 1964.

Yule, G. U., and M. G. Kendall: *An Introduction to the Theory of Statistics*, 14th ed., New York: Hafner Publishing Company, Inc., 1950.

A CATALOG OF SELECTED

DOVER BOOKS

IN SCIENCE AND MATHEMATICS

RELATIVITY, THERMODYNAMICS AND COSMOLOGY, Richard C. Tolman. Landmark study extends thermodynamics to special, general relativity; also applications of relativistic mechanics, thermodynamics to cosmological models. 501pp. 5⅜ × 8½. 65383-8 Pa. $12.95

APPLIED ANALYSIS, Cornelius Lanczos. Classic work on analysis and design of finite processes for approximating solution of analytical problems. Algebraic equations, matrices, harmonic analysis, quadrature methods, much more. 559pp. 5⅜ × 8½. 65656-X Pa. $12.95

SPECIAL RELATIVITY FOR PHYSICISTS, G. Stephenson and C.W. Kilmister. Concise elegant account for nonspecialists. Lorentz transformation, optical and dynamical applications, more. Bibliography. 108pp. 5⅜ × 8½. 65519-9 Pa. $4.95

INTRODUCTION TO ANALYSIS, Maxwell Rosenlicht. Unusually clear, accessible coverage of set theory, real number system, metric spaces, continuous functions, Riemann integration, multiple integrals, more. Wide range of problems. Undergraduate level. Bibliography. 254pp. 5⅜ × 8½. 65038-3 Pa. $7.95

INTRODUCTION TO QUANTUM MECHANICS With Applications to Chemistry, Linus Pauling & E. Bright Wilson, Jr. Classic undergraduate text by Nobel Prize winner applies quantum mechanics to chemical and physical problems. Numerous tables and figures enhance the text. Chapter bibliographies. Appendices. Index. 468pp. 5⅜ × 8½. 64871-0 Pa. $11.95

ASYMPTOTIC EXPANSIONS OF INTEGRALS, Norman Bleistein & Richard A. Handelsman. Best introduction to important field with applications in a variety of scientific disciplines. New preface. Problems. Diagrams. Tables. Bibliography. Index. 448pp. 5⅜ × 8½. 65082-0 Pa. $11.95

MATHEMATICS APPLIED TO CONTINUUM MECHANICS, Lee A. Segel. Analyzes models of fluid flow and solid deformation. For upper-level math, science and engineering students. 608pp. 5⅜ × 8½. 65369-2 Pa. $13.95

ELEMENTS OF REAL ANALYSIS, David A. Sprecher. Classic text covers fundamental concepts, real number system, point sets, functions of a real variable, Fourier series, much more. Over 500 exercises. 352pp. 5⅜ × 8½. 65385-4 Pa. $9.95

PHYSICAL PRINCIPLES OF THE QUANTUM THEORY, Werner Heisenberg. Nobel Laureate discusses quantum theory, uncertainty, wave mechanics, work of Dirac, Schroedinger, Compton, Wilson, Einstein, etc. 184pp. 5⅜ × 8½. 60113-7 Pa. $4.95

INTRODUCTORY REAL ANALYSIS, A.N. Kolmogorov, S.V. Fomin. Translated by Richard A. Silverman. Self-contained, evenly paced introduction to real and functional analysis. Some 350 problems. 403pp. 5⅜ × 8½. 61226-0 Pa. $9.95

PROBLEMS AND SOLUTIONS IN QUANTUM CHEMISTRY AND PHYSICS, Charles S. Johnson, Jr. and Lee G. Pedersen. Unusually varied problems, detailed solutions in coverage of quantum mechanics, wave mechanics, angular momentum, molecular spectroscopy, scattering theory, more. 280 problems plus 139 supplementary exercises. 430pp. 6½ × 9¼. 65236-X Pa. $11.95

ROTARY-WING AERODYNAMICS, W.Z. Stepniewski. Clear, concise text covers aerodynamic phenomena of the rotor and offers guidelines for helicopter performance evaluation. Originally prepared for NASA. 537 figures. 640pp. 6⅛ × 9¼.
64647-5 Pa. $14.95

DIFFERENTIAL GEOMETRY, Heinrich W. Guggenheimer. Local differential geometry as an application of advanced calculus and linear algebra. Curvature, transformation groups, surfaces, more. Exercises. 62 figures. 378pp. 5⅜ × 8½.
63433-7 Pa. $7.95

INTRODUCTION TO SPACE DYNAMICS, William Tyrrell Thomson. Comprehensive, classic introduction to space-flight engineering for advanced undergraduate and graduate students. Includes vector algebra, kinematics, transformation of coordinates. Bibliography. Index. 352pp. 5⅜ × 8½. 65113-4 Pa. $8.95

A SURVEY OF MINIMAL SURFACES, Robert Osserman. Up-to-date, in-depth discussion of the field for advanced students. Corrected and enlarged edition covers new developments. Includes numerous problems. 192pp. 5⅜ × 8½.
64998-9 Pa. $8.95

ANALYTICAL MECHANICS OF GEARS, Earle Buckingham. Indispensable reference for modern gear manufacture covers conjugate gear-tooth action, gear-tooth profiles of various gears, many other topics. 263 figures. 102 tables. 546pp. 5⅜ × 8½. 65712-4 Pa. $11.95

SET THEORY AND LOGIC, Robert R. Stoll. Lucid introduction to unified theory of mathematical concepts. Set theory and logic seen as tools for conceptual understanding of real number system. 496pp. 5⅜ × 8¼. 63829-4 Pa. $10.95

A HISTORY OF MECHANICS, René Dugas. Monumental study of mechanical principles from antiquity to quantum mechanics. Contributions of ancient Greeks, Galileo, Leonardo, Kepler, Lagrange, many others. 671pp. 5⅜ × 8½.
65632-2 Pa. $14.95

FAMOUS PROBLEMS OF GEOMETRY AND HOW TO SOLVE THEM, Benjamin Bold. Squaring the circle, trisecting the angle, duplicating the cube: learn their history, why they are impossible to solve, then solve them yourself. 128pp. 5⅜ × 8½. 24297-8 Pa. $3.95

MECHANICAL VIBRATIONS, J.P. Den Hartog. Classic textbook offers lucid explanations and illustrative models, applying theories of vibrations to a variety of practical industrial engineering problems. Numerous figures. 233 problems, solutions. Appendix. Index. Preface. 436pp. 5⅜ × 8½. 64785-4 Pa. $9.95

CURVATURE AND HOMOLOGY, Samuel I. Goldberg. Thorough treatment of specialized branch of differential geometry. Covers Riemannian manifolds, topology of differentiable manifolds, compact Lie groups, other topics. Exercises. 315pp. 5⅜ × 8½. 64314-X Pa. $8.95

HISTORY OF STRENGTH OF MATERIALS, Stephen P. Timoshenko. Excellent historical survey of the strength of materials with many references to the theories of elasticity and structure. 245 figures. 452pp. 5⅜ × 8½. 61187-6 Pa. $10.95